# Risk Engineering

**Series editor**

Dirk Proske, Bern University of Applied Sciences, Burgdorf, Bern, Switzerland

The Springer Book Series Risk Engineering can be considered as a starting point, looking from different views at Risks in Science, Engineering and Society. The book series publishes intense and detailed discussions of the various types of risks, causalities and risk assessment procedures. Although the book series is rooted in engineering, it goes beyond the thematic limitation, since decisions related to risks are never based on technical information alone. Therefore issues of "perceived safety and security" or "risk judgment" are compulsory when discussing technical risks, natural hazards, (environmental) health and social risks. One may argue that social risks are not related to technical risks, however it is well known that social risks are the highest risks for humans and are therefore immanent in all risk trade-offs. The book series tries to cover the discussion of all aspects of risks, hereby crossing the borders of scientific areas.

More information about this series at http://www.springer.com/series/11582

Martina Raue · Bernhard Streicher ·
Eva Lermer
Editors

# Perceived Safety

## A Multidisciplinary Perspective

Springer

*Editors*
Martina Raue
AgeLab
Massachusetts Institute of Technology
Cambridge, MA, USA

Eva Lermer
Department of Experimental Psychology
University of Regensburg
Regensburg, Bavaria, Germany

FOM University of Applied Sciences for
Economics and Management
Munich, Bayern, Germany

Bernhard Streicher
Institute of Psychology—RiskLab
UMIT—Private University for Health
Sciences, Medical Informatics and
Technology
Hall in Tyrol, Tyrol, Austria

ISSN 2195-433X ISSN 2195-4348 (electronic)
Risk Engineering
ISBN 978-3-030-11454-1 ISBN 978-3-030-11456-5 (eBook)
https://doi.org/10.1007/978-3-030-11456-5

Library of Congress Control Number: 2018967727



This Springer imprint is published by the registered company Springer Nature Switzerland AG
The registered company address is: Gewerbestrasse 11, 6330 Cham, Switzerland

To Niels

# Preface

Humans are inherently curious—from the very start of our lives, we strive to explore our environment. Children want to know what happens when they drop their plate or whether a pen can draw on the wall. As adults, we are in constant search for information about the world around us. This immanent curiosity is a driving force of human, social, and technological development. Innovation involves taking risks, but while being inherently curious, humans have a tendency toward risk avoidance. As a result, curiosity alone often does not induce progress because humans are also driven by a need for safety, control, and predictability.

Being curious and valuing safety are two sides of the same coin when it comes to taking risks. Thus, finding a balance between these two drivers of motivation is a major challenge for individuals, organizations, and societies. Researchers of various disciplines have extensively studied risk and uncertainty, but also implicitly acknowledged the role of certainty as a counter-pole. However, safety as a human need and motive has received considerably less attention than risk in academic research. On the one hand, life in western societies has become increasingly safe due to better standards in public health and policy, which have decreased risks for everyone. Accordingly, products and services concerning safety are considered one of the biggest industries in the world. On the other hand, our modern world often merely suggests safety or attempts to create a sense of safety when, in fact, complete safety is unattainable. The marketing industry, for example, advertises cars with innovative technologies that make driving safer than ever before—but driving remains one of the riskiest activities we engage in. Thus, the term safety often reflects a reduction in risk rather than the absence of harm and addresses the human need to *feel* safe. This volume aims to gain a better understanding of the role safety plays in our society and various disciplines, how humans perceive risks but also need safety, and how this understanding can lead to a culture that better supports people in their decision-making processes.

In this volume, researchers in engineering, philosophy, and psychology shed some light on the mechanisms of safety. In Chap. 1, Dirk Proske defines of terms related to safety from an engineering perspective and discusses whether optimal safety can be achieved. In Chap. 2, Dirk Proske discusses the categorization of risks and safety and the limits of such categorizations. In Chap. 3, Niels Gottschalk-Mazouz takes a philosophical perspective on terminology related to safety and introduces the term risk culture, thereby discussing rational and moral aspects of risk-taking and risk governance. In Chap. 4, Eric Eller and Dieter Frey take a psychological perspective by considering basic human needs and social determinants of perceived safety. In Chap. 5, Martina Raue and Elisabeth Schneider also take a psychological perspective and focus on human decision-making strategies. In Chap. 6, Eva Lermer, Bernhard Streicher, Martina Raue, and Dieter Frey shed some light on different factors underlying the assessment of risk. Chapters 7, 8, and 9 have a more applied focus. In Chap. 7, Susanne Gaube, Eva Lermer, and Peter Fischer discuss the relationship between risk perception and health-related behavior, while in Chap. 8, Robert Mauro addresses conflicts between facts and fears in aviation. Finally, risk sports are the focus of Chap. 9, in which Martina Raue, Bernhard Streicher, Eva Lermer, and Dieter Frey introduce several studies on the influence of physical activity on risk perception.

We would like to thank all of the authors for their valuable contributions in making this volume a multifaceted work that crosses disciplinary borders.

Cambridge, USA — Martina Raue
Hall in Tyrol, Austria — Bernhard Streicher
Regensburg/Munich, Germany — Eva Lermer

# Contents

Theoretical Aspects of Patient Safety

What is "Safety" and Is There "Optimal Safety"

# Editors and Contributors

## About the Editors

**Martina Raue** is a Research Scientist at the MIT AgeLab in Cambridge, Massachusetts, and an instructor in the Department of Psychology at Harvard University. She studies risk perception and decision-making over the life span, with a focus on the roles of time and emotion. She applies theories from social psychology to challenges and risks of longevity and examines ways to improve planning and preparing for later life. She received her Ph.D. in social psychology from the Ludwig Maximilian University Munich in Germany and her master's degree in psychology from the University of Basel in Switzerland.

**Bernhard Streicher** is a Full Professor of Social and Personality Psychology and Head of the Department of Psychology and Medical Sciences at the University for Health Sciences—UMIT in Hall, Tyrol in Austria. His research interests include the psychological mechanisms of decision-making under risk and uncertainty and the question of how to enhance the risk competencies of people and organizations. In addition to his scientific research, he works as a lecturer and consultant regarding the topic of 'risk' for profit and nonprofit organizations. He received his academic titles from the Ludwig Maximilian University Munich in Germany.

**Eva Lermer** is a Professor of Business Psychology at the FOM University of Applied Sciences for Economics and Management in Munich, Germany, and a Research Scientist in the Department of Experimental Psychology at the University of Regensburg, Germany. Her research interests include subjective risk assessment, decision-making, and de-biasing strategies. In addition to her scientific research, she works as consultant for profit and nonprofit organizations. She received her academic degrees from the University of Salzburg in Austria and the Ludwig Maximilian University Munich in Germany.

Martina Raue, Bernhard Streicher, and Eva Lermer jointly founded the *Risikolabor* (risk lab) at the Ludwig Maximilian University Munich in 2011. While currently based at different institutions, they continue to collaborate on various research projects investigating human perception of risk and influences on risk-taking behavior. In addition, they offer consulting and workshops on the topic. More about their work can be found at www.risikolabor.org.

## Contributors

**Eric Eller** Department of Psychology, Ludwig-Maximilian-University Munich, Munich, Germany

**Peter Fischer** Department of Experimental Psychology, University of Regensburg, Regensburg, Germany

**Dieter Frey** Department of Psychology, Ludwig-Maximilian-University Munich, Munich, Germany

**Susanne Gaube** Department of Experimental Psychology, University of Regensburg, Regensburg, Germany

**Niels Gottschalk-Mazouz** Department of Philosophy, University of Bayreuth, Bayreuth, Germany

**Eva Lermer** Department of Experimental Psychology, University of Regensburg, Regensburg, Germany
FOM University of Applied Sciences for Economics and Management, Munich, Germany

**Robert Mauro** Decision Research, Eugene, OR, USA
Psychology Department, University of Oregon, Eugene, OR, USA

**Dirk Proske** Bern University of Applied Sciences, Burgdorf, Switzerland

**Martina Raue** AgeLab, Massachusetts Institute of Technology, Cambridge, MA, USA

**Elisabeth Schneider** Department of Psychology, Ludwig Maximilian University of Munich Munich, Germany

**Bernhard Streicher** Institute of Psychology, UMIT—Private University for Health Sciences, Medical Informatics and Technology, Hall in Tyrol, Austria

# Part I
# Theoretical Aspects of Perceived Safety

# Chapter 1
# What Is "Safety" and Is There "Optimal Safety" in Engineering?

**Dirk Proske**

**Abstract** In this section a definition of the term "safety" based on freedom of resources is introduced. Such freedom of resources can also be used for the definition of the terms "danger" and "disaster". Additionally, the terms "safety", "danger" and "disaster" can be correlated to time horizons of planning. The introduced relationships will then be used for the discussion whether "optimal safety" is achievable or not. Currently, "optimal safety" is being intensively discussed in many disciplines such as the field of structural safety. Considering the definition of "safety", this paper will show that "optimal safety" is rather a theoretical issue and cannot be achieved under real world conditions. This statement fits very well not only to considerations in the field of system theory, but also to empirical observations. It is suggested that the term "optimal safety" is introduced as an assurance measure for engineers rather than for the public. As a solution the concept of integral risk management is introduced. One of the properties of this concept is the possibility of continuous improvement and therefore no optimal solution is claimed.

**Keywords** Safety · Risk · Optimal safety · Resources · Quality of life · Risk cycle

## 1.1 Introduction

### *1.1.1 Current Developments*

Over the last few years the question of optimal safety has been intensively discussed in many fields such as structural engineering. The question of optimal safety considers the selection of safety measures regarding minimum costs including

D. Proske (✉)
Bern University of Applied Sciences, Pestalozzistrasse 20, Postfach 1058, 3401 Burgdorf, Switzerland
e-mail: dirk.proske@bfh.chc



M. Raue et al. (eds.), *Perceived Safety*, Risk Engineering,
https://doi.org/10.1007/978-3-030-11456-5_1

failure costs. For example, building a weak and cheap construction which will fail and has to be re-build regularly or building a very strong and expensive structure which will remain for a long time without failure. This question of optimal safety is of particular interest for the development of general safety requirements related to all technical products, such as building structures, airplanes, cars etc. For example, within the last decades, the general safety concept in structural engineering has been updated from a simple global safety factor concept to a safety concept which is based on probabilistic issues and which is able to adequately consider such questions. Therefore, the update to the new safety concept initiated debates regarding the optimal safety of structures.

The question of optimal safety has been mainly answered through the economical optimization of the spending of resources. This includes the important and true consideration that resources for humans and societies are limited.

In structural engineering, usually the sum, the overall costs of the production cost and the cost of failure (disadvantages) are compared with the possible gains of creating such a structure (advantages). The combination of these two cost components as shown in Fig. 1.1 yield to an overall cost function with a minimum value according to some adaptable structural design parameters included. Such design parameters can be, for example, the strength of the building material or the geometries. This overall cost function is based on economic considerations. It is actually a cost-benefit analysis, or, how it may be called here: an advantage-disadvantage-analysis. The difference between an advantage-disadvantage-analysis and a cost-benefit-analysis is the inclusion of further advantages and disadvantages, which might not be directly presented as economic values. For example, sometimes additional measures such as those found within the quality of life parameters are incorporated. Dimensions of such factors are shown in Fig. 1.2 which provides a good impression regarding the diversity and the scale of such factors. The application of such quality of life parameters has a long tradition in medicine and has been applied in structural engineering for approximately two decades. For example, the Life Quality Index (LQI) by Nathwani et al. (1997) has become widely used in several engineering fields (Proske 2004; 2009).

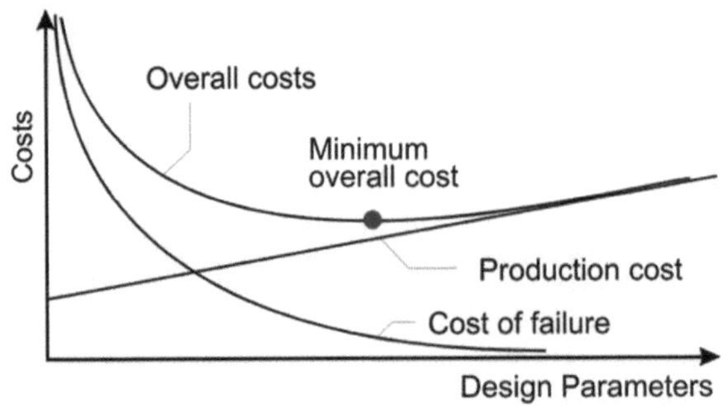


**Fig. 1.1** Widely used function of overall structural cost depending on several parameters

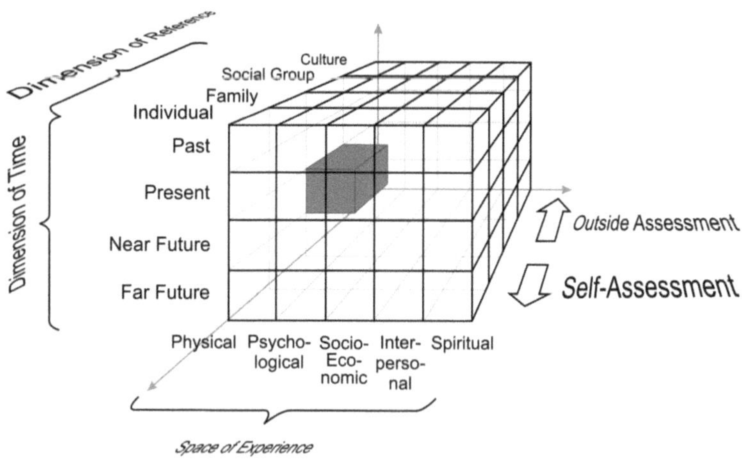


**Fig. 1.2** Dimension of quality of life according to Küchler and Schreiber (1989)

## 1.1.2 Limitation of the Current Developments

Although the search for performance measures as a basis of optimization procedures has in many fields yielded to the application of quality of life parameters, it does not necessarily mean that this strategy has been successful. It shows only that entirely pecuniary-based performance measures might be insufficient. If one considers for example the history of quality of life measures in medicine since 1948, one will find that now a huge variety of such parameters (up to 1500 according to Porzsolt and Rist 1997, Kaspar 2004) have been developed for very special applications. Such a specialization requires major assumptions inside the parameters. For example, the LQI assumes a trade-off between working time and leisure time for individuals. Although this might be true for some people, most people enjoy working (von Cube 1997) if the working conditions and the working content fit to personal preferences. The choice of using the average lifetime as a major indicator for damage has also been criticized (Proske 2004, but see also Müller 2002). The question, whether a quality of life parameter can be constructed on only a very limited number of parameters to be applicable still remains. Again, Fig. 1.2 should be mentioned as giving an impression about the dimensions of quality of life (Küchler and Schreiber 1989).

The comparison between the different dimensions and the simplified definition of the LQI makes limitations visible. For example, many psychological effects are not considered in the LQI. Since people are so strongly affected by their individual, social and cultural experience, these effects can rarely be excluded in useful quality of life measures and even further in decision-making processes. Many works have been done in this field such as Fischhoff et al. (1981), Slovic (1999), Covello (1991), Zwick and Renn (2002) or Schütz et al. (2003). For a general summary see Proske (2009).

Returning to the original question, the terms "safety" and "optimal safety" still need to be defined.

## 1.2 Terms

### *1.2.1 The Term "Safety"*

The term "safety" is often defined as a situation with a lower risk compared to an acceptable risk or a situation "without any danger impending" (Proske 2009). Other definitions describe safety as "peace of mind". Whereas the first definition that uses the term "risk" is already based on a substitution, the term "peace of mind" is a more general definition. The author considers "safety" to be the result of an evaluation process of a certain situation. The evaluation can be carried out by every system that is able to perform a decision-making process, such as animals, humans, societies or computers (which use algorithms). However, algorithms usually use some numerical representation. The following equation shows an example from a code of practice of defining safety $S$ when the existing risk $R$ is less than an allowable risk:

$$\begin{aligned} \text{existing}\, R &\leq \text{permitted}\, R \rightarrow S \\ \text{existing}\, R &> \text{permitted}\, R \rightarrow \$ \end{aligned}$$

Also, the author considers human feelings as a result of a decision-making process. Therefore, safety is understood here as a feeling; safety is a perception. Furthermore, the decision-making process deals with the question whether some resources have to be spent to decrease hazards and danger to an acceptable level or not, for example spending money for mitigation measures. In other terms "safety" is a feeling that no further resources have to be spent to decrease any threats. If one considers the term "no further resources have to be spent" as a degree of freedom of resources, one can define "safety" as a value of a function which includes the degree of freedom of resources. Furthermore, one can assume that the degree of freedom is related to some degree of distress and relaxation. Whereas in safe conditions relaxation occurs, in dangerous situations a high degree of distress is clearly reached.

The possible shape of the function between degree of relaxation, which ranges from "danger" to "peace of mind," and the value of the function as degree of freedom of resources is shown in Fig. 1.3. The degree of freedom of resources describes the extent to which a person or a society can decide on the use of its own resources independently of external influences. It is assumed here that the relationship is non-linear with at least one region of over-proportional growth of the relative freedom of resources. In Fig. 1.3 this region of over-proportional growth is defined as the starting point of the safety region:

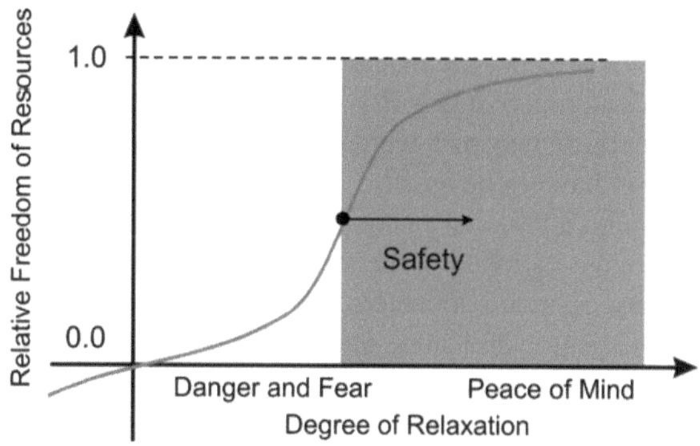


**Fig. 1.3** Assumed function shape between degree of relaxation and relative freedom of resources

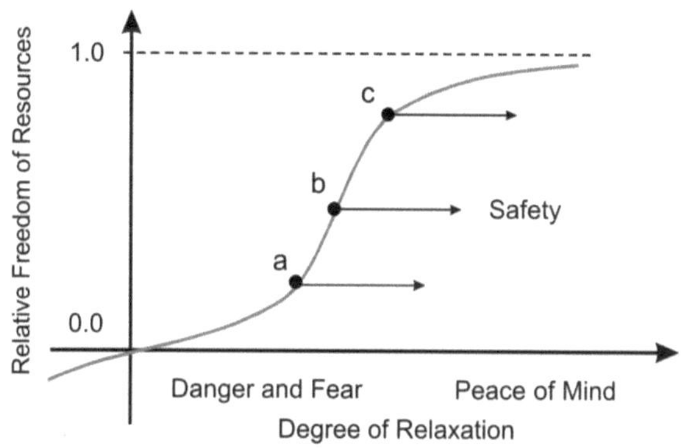


**Fig. 1.4** Assumed function shape between degree of relaxation and relative freedom of resources with different starting points of the safety region

$$S = \{x | f''(x) = 0\}$$

However, the question still remains, where the region of safety starts since other points are possible. Additionally, the selection of this point may be highly individual. In Fig. 1.4 further points are shown considering either regions of maximum curvature or the point of inflection.

The degree of relaxation (*DoR*) or the "perceived safety" can be evaluated based on a mathematical function considering the input variables:

$$DoR = f(a, b, c, d\ldots)$$

Furthermore, the influence parameters for the degree of relaxation have to be chosen. As already mentioned, safety is understood here as a feeling. By definition, this is a subjective evaluation of a situation and therefore the term "perceived safety" is actually a pleonasm. Often the terms "subjective risk judgment" and "subjective safety assessment" are used as well. However, the term "perceived safety" has become very popular in scientific literature and shall be used here. The term perceived safety considers subjective effects, for example trust. Covello and colleagues (2001) have stated that trust might shift the individual acceptable risk by a factor of 2000. That means, if one convinces people through dialogue that a house is safe, a much higher risk (no resources are spent) will be accepted, whereas with only a few negative words trust can be destroyed and further resources for protection have to be spent.

Many additional factors, such as voluntariness, benefit, control, age and experience are included in the term perceived safety (Proske 2009, Covello 1991). The variety of such parameters shows that the mathematical-theoretical formulation of such a degree of distress and relaxation is limited, in other terms human behavior is too complex to be explained by a simple mathematical formula. Therefore, input data is most frequently provided by surveys.

Incidentally, the introduced definition of safety also gives the opportunity to define a relationship between the terms "disaster", "danger" and "safety" and the freedom of resources (Fig. 1.5). The discussion on safety has already introduced danger as a situation, where the majority of resources are spent to re-establish the condition of safety (= not spending resources). Under an extreme situation of danger, no freedom of resources exists anymore; since all resources are spent to re-establish safety. The term disaster then describes a circumstance, where the resources are overloaded (negative). Here external resources, such as help from other persons, other countries etc., are required to re-establish safety. This indeed fits very well to common definitions of disaster stating, that external help is required (Proske 2009).

Additionally, the introduced definition can be transferred to the time scale of planning and spending resources, as shown in Fig. 1.6. The time horizon of planning alters dramatically in correlation with the states of danger and safety. Under the state of safety and peace of mind the time horizon shows a great diversity of planning times ranging from zero (present) to decades or even longer. In emergency states, the time horizon only considers very short time durations, such as seconds or minutes.

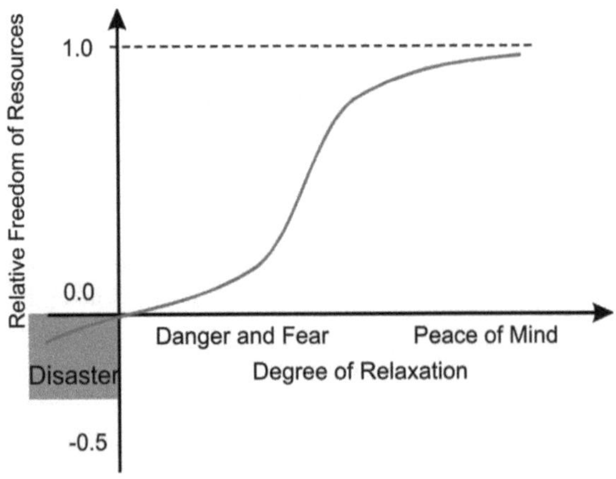


**Fig. 1.5** Assumed function shape between degree of distress and relative freedom of resources with the definition of a disaster region with negative resources (the need for external help)

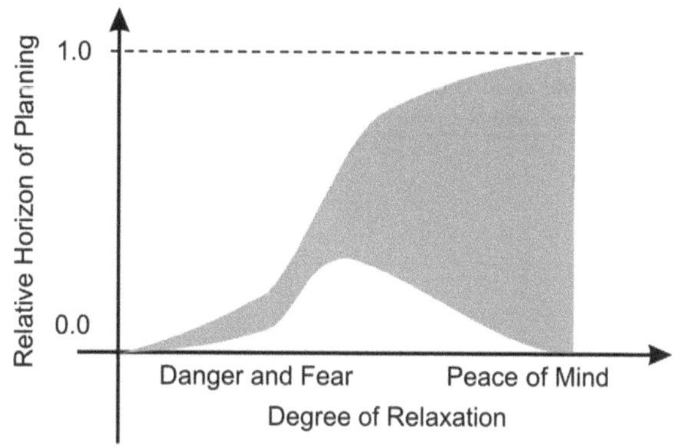


**Fig. 1.6** Assumed relationship between degree of distress and relative horizon of planning

It is necessary now to return to the term "optimal safety". This term is mainly applied as efficiency criteria to reach a maximum of utility. Mostly the Pareto criteria or the Kaldor-Hicks compensation tests are used (Pliefke and Peil 2007). However, here instead "optimal safety" is defined as a condition, which yields to a maximum performance of humans. Such maximum performance is described in relation to different degrees of stress and relaxation by the Yerkes-Dodson-curve as shown in Fig. 1.7 (Proske 2009).

Figure 1.7 indicates that the majority of humans do not reach their maximum work performance under extreme safe conditions and high degrees of freedom of resources respectively. Instead, humans tend to return to unsafe regions reaching for further gains (Evans 1986), as shown in Fig. 1.8. The author considers the ways, in which humans return to such stages are manifold. For example, people may show learned helplessness behavior or they may show homeostatis of risks as observed with ABS systems in cars, were people deliberately rely on such a system by driving riskier (Proske 2009).

In general, humans follow non-linear utility functions as shown in Fig. 1.9 with high amounts of subjective elements in the evaluation process of the utility.

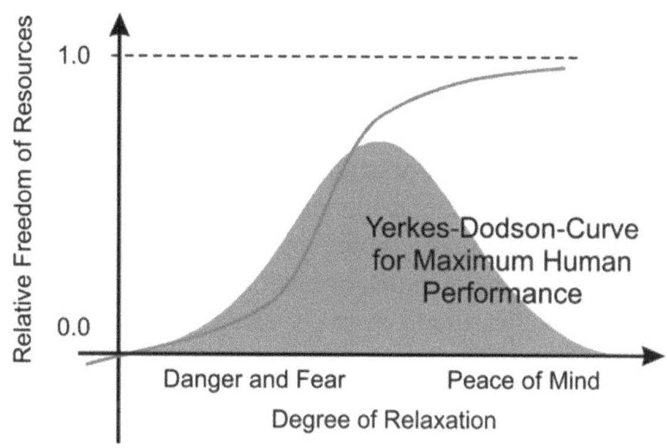


**Fig. 1.7** Assumed function shape between degree of distress and relative freedom of resources with the Yerkes-Dodson-curve as a relationship between human performance and degree of distress

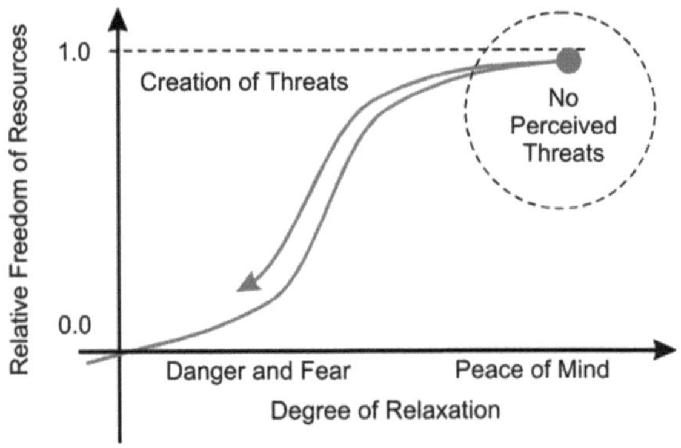


**Fig. 1.8** Assumed function shape between degree of distress and relative freedom of resources and the return curvature

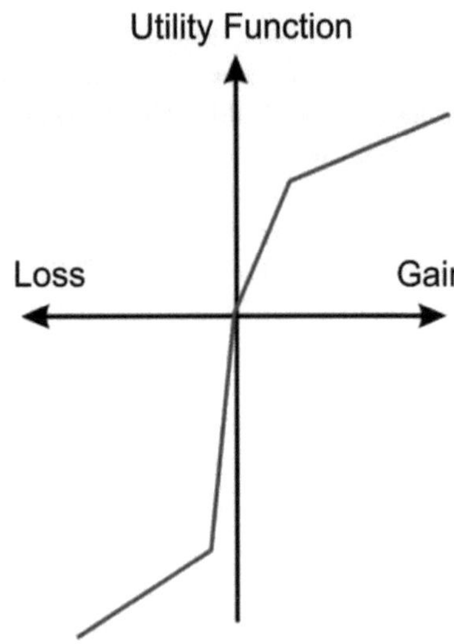


**Fig. 1.9** Utility function of gain and loss according to Kahneman and Tversky (1981)

In scientific works in beta sciences (e.g. natural sciences, engineering sciences) subjective elements are often neglected. However, decisions under real world conditions include such effects. The question remains, if "optimal safety" regions based on some beta scientific investigations indeed represent an "optimal safety".

Subjective judgment often considers many elements, which seem to have only a small correlation to the relevant indicator. However, the wideness of the parameter is extremely high. For example, someone does not sign a working contract based on some astrological consideration. In contrast, scientific approaches mainly consider elements with high correlation to be a relevant indicator (Fig. 1.10). For scientific approaches major indetermination in nature may limit the scientific approach, as scale cascades found in many examples heavily influence such approaches. Scale effects are effects which yield to changing relevant indicators based on the size of the problem.

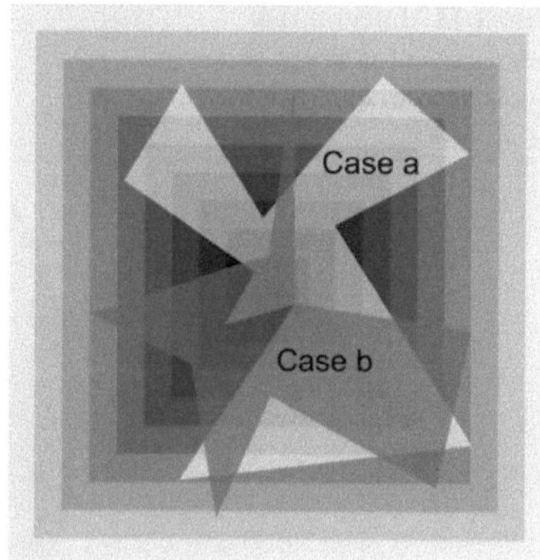


**Fig. 1.10** Visualized correlation matrix of possible input variables for an investigation (rectangles) with increasing correlation shown by increasing darkness and the choice of the parameters for an optimization procedure based on (a) a subjective evaluation and (b) a rational model

Many works have shown the limitation of optimization processes not only for safety, but also for other applications (Buxmann 1998). The following remarks from the mathematical research project entitled "Robust mathematical modeling" give the same results (Société de Calcul Mathématique 2017):

1. There is no such thing, in real life, as a precise problem. As we already saw, the objectives are usually uncertain, the laws are vague and data is missing. If you take a general problem and make it precise, you always make it precise in the wrong way or, if your description is correct now, it will not be tomorrow, because some things will have changed.
2. If you bring a precise answer, it seems to indicate that the problem was exactly this one, which is not the case. The precision of the answer is a wrong indication of the precision of the question. There is now a dishonest dissimulation of the true nature of the problem.

Piecha (1999) states that flowers in a room may not be considered as the optimal functional design of decors in offices; however, Piecha has shown that over the long-term small disturbances such as ones created by flowers might increase efficiency of the workers. Furthermore, as Arrow (1951) has shown no procedures exist to find optimal solutions for some types of systems. See here also Proske (2006) and Riedl (2000).

### *1.2.2 Solution*

If initial values and the criteria for "optimal safety" cannot be defined, one should dismiss the concept of "optimal safety", since there is only "optimal safety" for certain criteria given and some initial values chosen. How strong the relation to the real-world problem is, remains to be proven.

On the other hand, an improvement of current safety conditions should be gained. Here concepts of risk-informed decisions (Arrow et al. 1996), integral risk

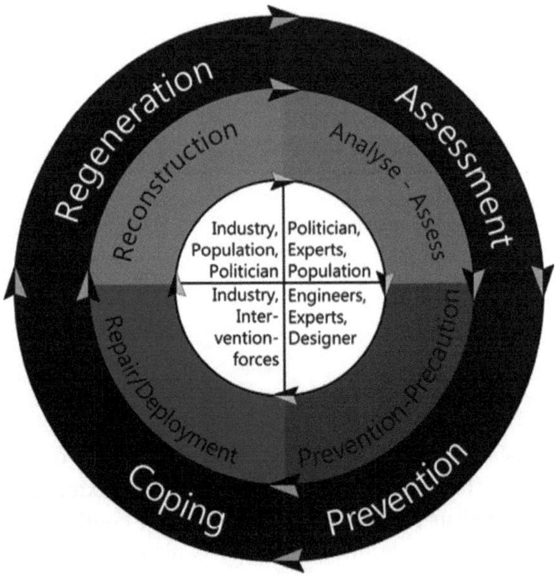


**Fig. 1.11** Integral risk management concept as a cycle according to Kienholz et al. (2004)

management (Kienholz et al. 2004), lifecycle management, living Probabilistic Safety Analysis or the regular re-evaluation of the "perceived safety" could be helpful. These approaches do not promise an optimal safety, but they promise permanent improvement (Fig. 1.11). These concepts are much more realistic compared to the concept of "optimal safety".

## References


Arrow, K. J. (1951). *Social choice and individual values.* New York: Wiley.

Arrow, K. J., et al. (1996). Is there a role for benefit-cost analysis in environmental, health and safety regulations. *Science, 272,* 221–222.

Buxmann, O. (1998). Erwiderungsvortrag von Herrn Prof. Dr.-Ing. Otto Buxmann zur Ehrenpromotion: Ursachen für das mechanische Verhalten von Konstruktionen bei Unfällen. *Wissenschaftliche Zeitschrift der Technischen Universität Dresden, 47*(5–6), 145–147.

Covello, V. T. (1991). Risk comparisons and risk communications: Issues and problems in comparing health and environmental risks. In R. E. Kasperson (Ed.), *Communicating risk to the public* (pp. 79–124). Dordrecht: Kluwer Academic Publishers.

Covello, V. T., Peters, R. G., Wojtecki, J. G., & Hyde, R. C. (2001). Risk communication, the west Nile virus epidemic, and bioterrorism: Responding to the communication challenges posed by the intentional or unintentional release of a pathogen in an urban setting. *Journal of Urban Health: Bulletin of the New York Academy of Medicine, 78*(2), 382–391.

Fischhoff, B., Lichtenstein, S., Slovic, P., Derby, S. L., & Keeney, R. L. (1981). *Acceptable risk.* Cambridge: Cambridge University Press.

Evans, L. (1986). Risk homeostasis theory and traffic accident data. *Risk Analysis, 6*(1), 81–94.

Kaspar, T. (2004). *Klinisch-somatische Parameter in der Therapie schwerer Herzinsuffizienz unter besonderer Berücksichtigung der gesundheitsbezogenen Lebensqualität Verlaufsanalyse des*

*Einflusses klinisch-somatischer Parameter auf die Lebensqualität.* München: Ludwig-Maximilians-Universität.
Kienholz, H., Krummenacher, B., Kipfer, A., & Perret, S. (2004). Aspect of integral risk management in practice–Considerations. *Österreichische Wasser- und Abfallwirtschaft, 56*(3–4), 43–50.
Küchler, T., & Schreiber, H. W. (1989). Lebensqualität in der Allgemeinchirurgie - Konzepte und praktische Möglichkeiten der Messung. *Hamburger Ärzteblatt, 43,* 246–250.
Müller, U. (2002). Quality of life at its end: Assessment by patients and doctors. In A. Gimmler, C. Lenk, & G. Aumüller (Eds.), *Health and quality of life.* Hamburg, Berlin: LIT-Verlag Münster.
Nathwani, J. S., Lind, N. C., & Pandey, M. D. (1997). *Affordable safety by choice: The life quality method.* Ontario, Canada: Institute of Risk Research, University of Waterloo.
Piecha, A. (1999). Die Begründbarkeit ästhetischer Werturteile. Fachbereich: Kultur- u. Geowissenschaften, Universität Osnabrück, Dissertation.
Pliefke, T., & Peil, U. (2007). On the integration of equality considerations into the life quality concept for managing disaster risk. In L. Taerwe & D. Proske (Eds.), *Proceedings of the 5th International Probabilistic Workshop*, Ghent, pp. 267–282.
Porzsolt, F., & Rist, C. (1997). Lebensqualitätsforschung in der Onkologie: Instrumente und Anwendung. In H. Bullinger (Ed.), *Lebensqualitätsforschung* (pp. 19–21). Stuttgart, New York: Schattauer.
Proske, D. (2004). *Katalog der Risiken.* Dresden: Dirk Proske Verlag.
Proske, D. (2006). *Unbestimmte Welt.* Dresden, Wien: Dirk Proske Verlag.
Proske, D. (2009). *Catalogue of risks.* Heidelberg, Berlin, New York: Springer.
Riedl, R. (2000). *Strukturen der Komplexität – Eine Morphologie des Erkennens und Erklärens.* Heidelberg: Springer Verlag.
Schütz, H., Wiedemann, P., Hennings, W., Mertens J., & Clauberg, M. (2003). Vergleichende Risikobewertung – Konzepte, Probleme und Anwendungsmöglichkeiten, Abschlussbericht zum BfS Projekt StSch 4217. Forschungsgruppe Jülich GmbH – Programmgruppe Mensch, Umwelt, Technik.
Slovic, P. (1999). Trust, emotion, sex, politics, and science: Surveying the risk-assessment battlefield. *Risk Analysis, 19*(4), 689–701.
Société de Calcul Mathématique. (2017). Robust mathematical modeling. http://www.scmsa.com/RMM/robust3.htm.
Tversky, A., & Kahneman, D. (1981). The framing of decisions and the psychology of choice. *Science, 211,* 453–489.
von Cube, F. (1997). *Fordern statt Verwöhnen.* München: Piper Verlag GmbH.
Zwick, M. M., & Renn, O. (2002). *Wahrnehmung und Bewertung von Risiken*: Ergebnisse des Risikosurvey Baden-Württemberg 2001. Nr. 202, Mai 2002, Arbeitsbericht der Akademie für Technikfolgenabschätzung und der Universität Stuttgart, Lehrstuhl für Technik- und Umweltsoziologie.

# Chapter 2
# Categorization of Safety and Risk

**Dirk Proske**

**Abstract** Risks can be categorized and ranked based on different characteristics. Such characteristics can be causes of damages or consequences or magnitudes of risks. Categorizations may improve the understanding of risks and, even more importantly, may help decision-makers to deal properly with risks in terms of risk-informed decisions. However, it has often been shown that the categorization of risks cannot be completed because the definition of risks is challenging and the size of the risks depends strongly on the choice of risk parameter and context. For example, many statistical investigations have shown that health risks are of utmost importance for human lives since 95% of all deaths in developed countries are related to health problems. Other risk studies reveal that the greatest risks to humans are social failures such as war or unemployment since many health issues are related to social failures. Such categorizations are even more difficult for emerging risks related to new technologies or current changes in social systems. For such systems, experience and therefore statistical data is still missing. Additionally, some of these systems belong to the class of complex systems with unknown causal chains.

**Keywords** Safety · Risk · Categorization

## 2.1 Introduction

A definition of safety was given in the chapter "What is "safety" and is there "optimal safety" in engineering?" as a condition with no need to spend resources or to take actions related to a certain internal or external event or process. This definition shows that safety is a general requirement for humans since we always have to deal with limited resources such as limited time or limited financial

D. Proske (✉)
Bern University of Applied Sciences, Pestalozzistrasse 20, Postfach 1058,
3401 Burgdorf, Switzerland
e-mail: dirk.proske@bfh.chc



M. Raue et al. (eds.), *Perceived Safety*, Risk Engineering,
https://doi.org/10.1007/978-3-030-11456-5_2

resources and are unable to deal with an unlimited number of problems at the same time. Therefore, certain processes have to be safe, at least for a certain time, for example breathing, drinking, eating, sleeping and working. In modern societies with their high complexity and their diversity of issues, safety remains an essential part of human existence and features legal, technical, economic, psychological, social and cultural properties. For example, the term "safety" appears more than 20 times in the Swiss constitution, showing the importance of this term in political discussions, but it lacks a definition.

In general, each scientific area uses different definitions of safety. For example, in technical fields either safety factors, probabilities of failure or various risk values are used. It is impossible to cover all definitions and categorizations in this chapter; however, some indications and examples will be given. This section focuses mainly on the differences in the scientific areas. Again, as an example, people consider different thresholds for the need of action or the amount of resources spent depending on who is responsible for the potential damage process (e.g. the risk). The public requires state authority actions in cases, in which a private person does not see any need for action. A very nice example for this discrepancy is speed limit, which to a certain extend decreases risks on the road system but is violated by a significant number of individuals. Obviously, such key elements have to and can be used to categorize risk.

## 2.2 Theory of Categorizations

In virtually all scientific areas, rules and recommendations exist to build categorizations of subjects, such as creatures, elementary particles, chemicals and others. A well-chosen categorization or order system provides an easy, comfortable way to group items subject to certain properties. In general, definitions and categorizations improve communication. Some categorization systems are partially based on historical developments, such as the categorization of animals or plants. However, common properties of the objects, such as inherent properties (mammals, system of elements) are the main source of categorizations and definitions.

Theoretically, a categorization is the application of assignment rules to an object to assign the object to a certain group of objects. In a first step, therefore, the assignment rules for risks must be developed.

Assignment rules for risks can be based on the causal origin of the risks, on the magnitude or other properties of the consequences, on the frequency of the risks or on other properties of the risks such as level of dread and level of the unknown (Slovic et al. 1980). The author has used the categorization based on the origin of the risks in Proske (2009). As an example, a categorization of risks based on the type of consequences can consider the sheer amount of consequences, the temporal sequence of consequences as shown in Fig. 2.1 as function of the number of casualties over time or the type of consequences such as fatalities, harm, financial damage etc. The function for flooding includes various effects, such as the duration

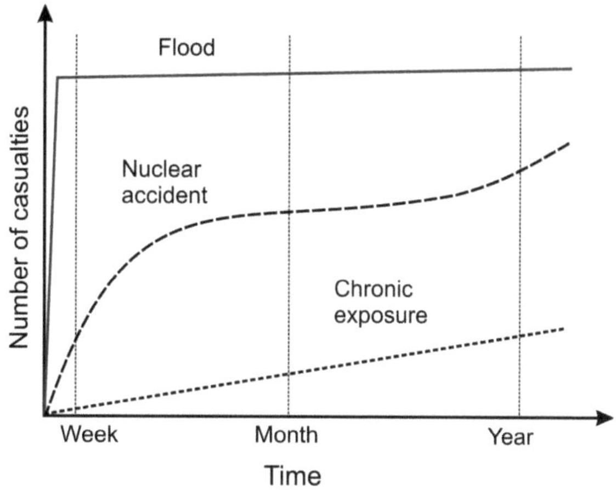


**Fig. 2.1** Function of the number of casualties over time related to different types of accidents (Dombrowski 2006)

of a flood and the resulting destruction of houses, food shortages and diseases such as during and after the Banqiao dam failure in 1975. The function for any chronic exposure shows a simple linear model. The assignment rules of risk categorizations have by no means to be restricted to rational measures. Common risk categorizations are based on subjective properties, such as perceived responsibilities or scariness (Slovic et al. 1980). Therefore, the risk categorization can be related to the risk assessment tools, or even better, to the perspective, under which the risk is assessed. Weichselgartner (2002) has grouped risks based on the following perspectives:

- engineering, technical, mathematical
- psychological, cognitive
- cultural, social
- geographical, natural space related.

Renn (1992) went even further considering the following risk perspectives:

- insurance, statistical
- toxicological, epidemiological
- engineering-technical
- economical
- psychological
- social
- cultural.

An extended version of the work by Renn (1992, 1998) is taken from Weichselgartner (2002) and shown in Fig. 2.2.

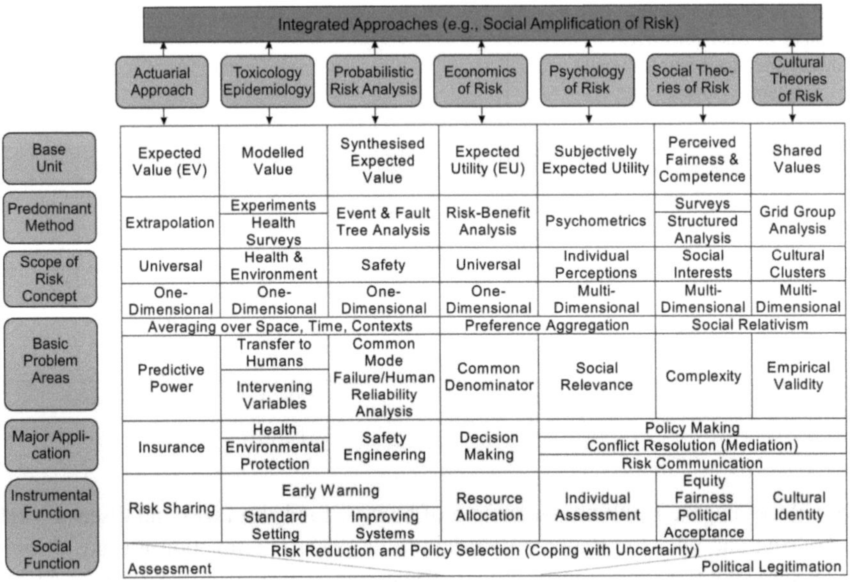


**Fig. 2.2** Categorization of risk perspectives according to Renn (1992, 1998) and Weichselgartner (2002)

## 2.3 Types of Safety and Risks

As mentioned, subjective elements are of significant importance for the categorization of risks. For example, people clearly distinguish between situations, in which they are responsible or in which they are not responsible but the society as a whole. In most cases people define higher requirements, if they are not responsible for the safety itself.

For example, one can consider the killing of people by earthquakes. Some scientists argue that not the earthquake itself is killing the people, but the collapsing structures and therefore people are killed by the failure of engineers. Another example is the re-settlement of great carnivores in Central Europe. Whereas the responsibility of the killing or severe injuries of humans caused by dogs in human settlements is officially related to the owner of the dog, the potential killing of humans by carnivores is often related to the inadequate behavior of the humans confronted with carnivores, not to the organizations responsible for the re-settlement. It should be mentioned, that the killings by dogs are substantially larger than the killings by great carnivores in Central Europe.

This inequality of responsibilities is seen over all types of disasters and risks.

In Fig. 2.3 each axis shows the results given by respondents whether a disaster was purely caused by natural effects yielding to factor 1.0 on the y-axis or it was a man-made disaster yielding to a factor of 1.0 on the x-axis. As the figure shows, the public sees none of the well-known natural hazards, such as floods or earthquakes,

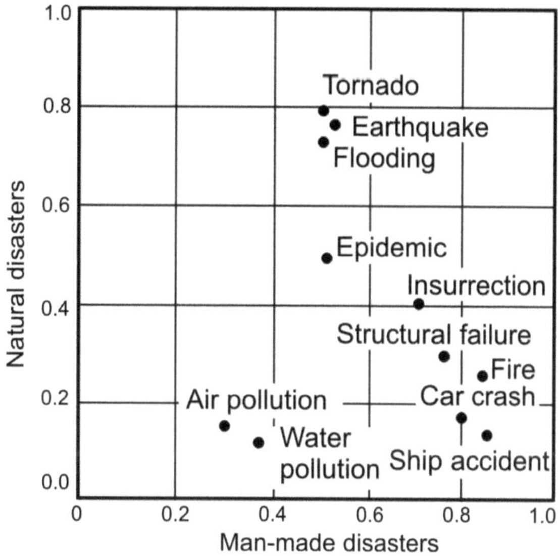


**Fig. 2.3** Causes of disasters and risk respectively according to a survey (Karger 1996)

as entirely natural disasters. According to the survey earthquakes or floods have 80% natural cause, but also a 50% man-made contribution. This can be directly related to the above-mentioned example of earthquake killings: people not only consider the earthquake as the initiators, they also blame the engineers for insufficient preparedness. Interestingly, the values do not sum up to 100% but exceed this theoretical maximum value. Therefore, subjective elements have to be considered in the categorization of risks (Proske 2009).

Several people besides the so far mentioned authors have strongly introduced subjective elements for the categorization of risks, such as Starr (1969), Tversky and Kahnemann (1974), Slovic et al. (1980), Fischhoff et al. (1981), Viscusi (1995), Wiedemann (1999) and Covello et al. (2001). Some of the approaches merge statistical risk values with subjective correction factors such as Covello et al. (2001). In contrast, other traditional statistical-engineering approaches express the limitations by the explicit definition of scenarios (Kaplan and Garrick 1981).

Klinke and Renn (1999) introduced a further risk categorization based on Greek-antic characters. This allows the identification of situations, e.g., the categorization in a very simple way. The concept is shown in relation to the traditional risk diagram in Fig. 2.4 and examples are given in Fig. 2.5.

All the categorizations of risks are finally used for proofs of safety and human wellbeing respectively, for example showing that observed risks are lower than acceptable risk. More general, human health and wellbeing is based on the compliance between observed or perceived values of various parameters and permitted values or ranges of such parameters; for further details see in Proske (2009) the section on "Quality of Life". The parameters belong to different scientific areas and are therefore called physical, chemical, biological, social, cultural and psychological acceptance criteria. Subsequently, risks can be categorised based on the scientific area.

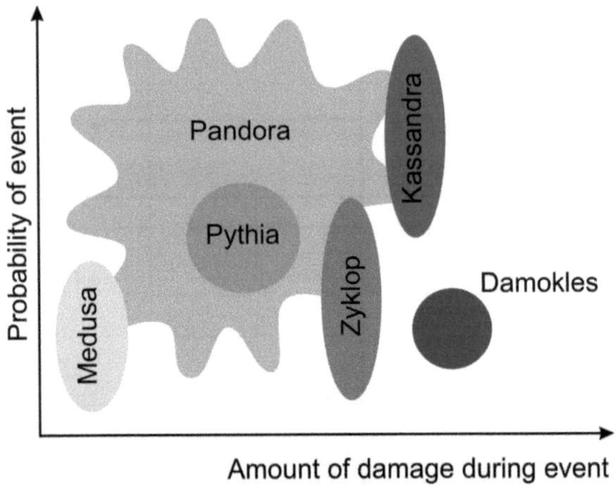


**Fig. 2.4** Categorization of risks based on Klinke & Renn (WGBU 1999)

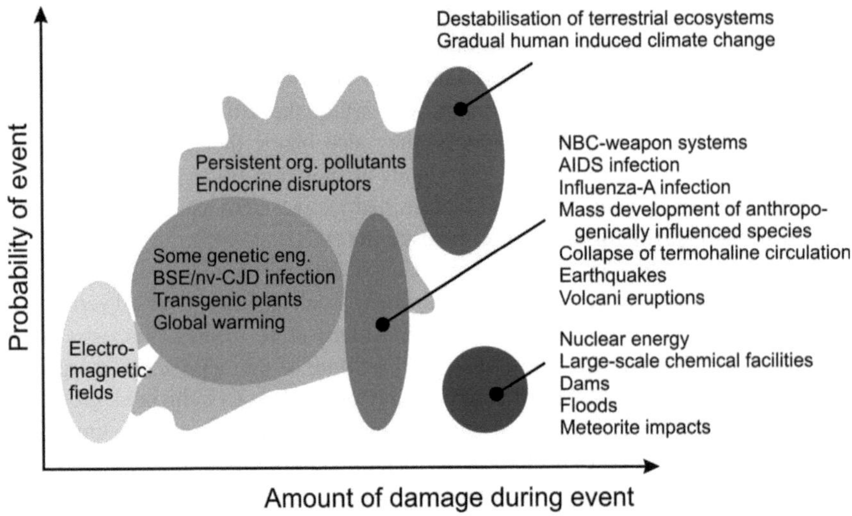


**Fig. 2.5** Categorization of risks based on Klinke & Renn (WGBU 1999)

In contrast, risks can be categorized to the parameter exceeded, for example limited oxygen provision. This would put the risks of drowning, suffocation or failure of the lung into one risk type. Another good example would be the damage of DNA, which can be related to chemicals, such as mushroom toxics, radioactive radiation, certain diseases or simply aging. What these examples already indicate, is the difficulty to consider effects from many different scientific fields. Since our educational and scientific system is based on specialisation, usually we try to understand different processes and consequences related to a certain scientific field.

From the point of view of science philosophy, this approach is correct because we move from the known to the unknown in small steps and thus avoid major mistakes. From the point of view of risk science, this approach may be limited since the public does not estimate and categorize risks related to fields of science, but rather in a heuristic approach, for example related to media attention. Sometimes the first approach is called bottom-up, where as an example technical details are considered to model risks, and the other approach is called top-down, where risks are considered from a general view without technical details. In theory, both approaches should yield to the same results, but often fail to do so.

Even further, all risk assessments are time dependent. Metzner (2002) has shown the time dependent risk development in a two-dimensional diagram considering the objective (e.g., mathematical-statistical) and the subjective risk evaluation. It becomes clear from Fig. 2.6, that risks from technologies, and also from natural disasters, which are not included in this diagram, show time-dependencies. This may yield to the effect, that safety measures against a risk, which reached a certain state in this diagram, are related to another risk, which is even larger, but not yet considered as larger. In other words, safety measures against risks can be placed on different states in the diagram and are therefore not considered equally as they should for decision making. Even worse, in many such cases safety measures are considered as including no risks at all.

Figure 2.7 tries to visualize this conclusion above in an example. In Fig. 2.7 a mitigation measure introduces a new risk c. However, most people consider only the decrease of the original risk from a to b. Risks a, b and c may belong to different risk classes or types, so one risk may be a technical one and the other one may be a social risk.

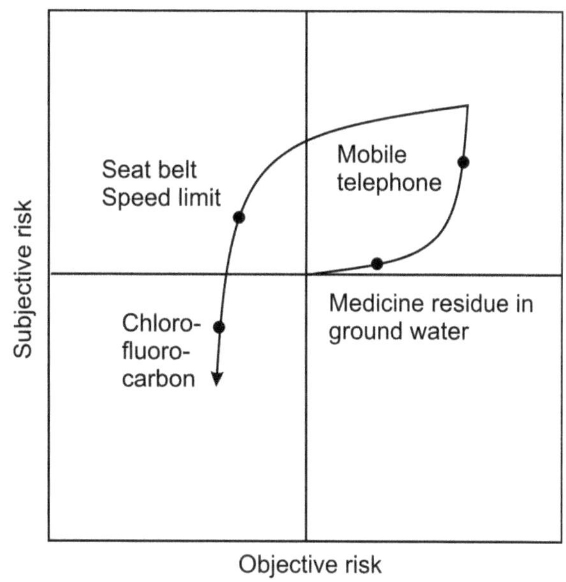


**Fig. 2.6** The function of objective (= mathematical-statistical risk expression) and subjective risk perception. The technologies move with time along the line shown (adapted from Metzner 2002)

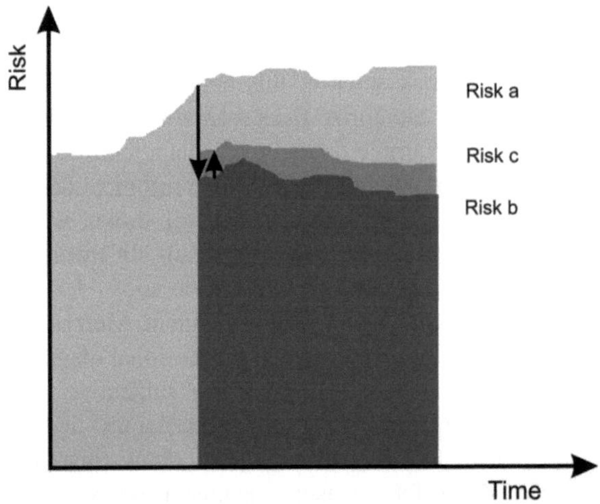


**Fig. 2.7** Decreasing a risk a to risk b by implying a safety measure imposing a new risk c

It is almost impossible to consider all risks adequately. The capabilities of all systems to categorize or rank all items under all conditions are limited. Therefore, systems, for example humans, develop simplified approaches. For example humans overestimate some new risks and underestimate some existing risks due to limited experience. This fact provides an explanation for the degradation of risk awareness over time for statistically constant risks. In other words, empirical data shows a constant risk but the subjective evaluation indicates a decreasing risk (see Fig. 2.8). On the other hand, a statistically changing risk may subjectively be seen as steady risk.

By accepting the limitations of risk categorization, we have to install another technique to provide safety for humans and societies. Such an approach is the concept of safety cultures. It is based on the assumption that disasters do not happen

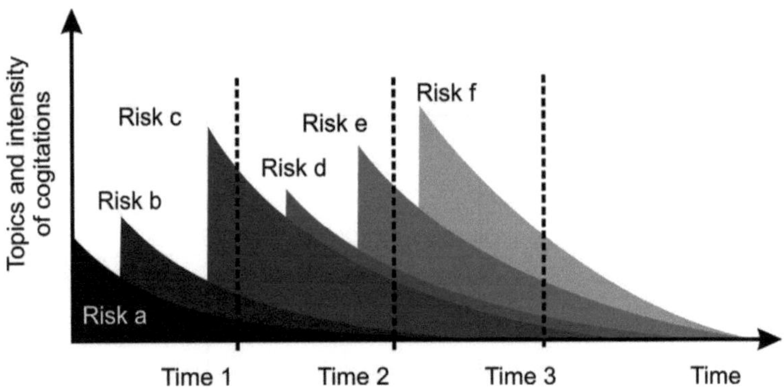


**Fig. 2.8** Awareness of various risks over time

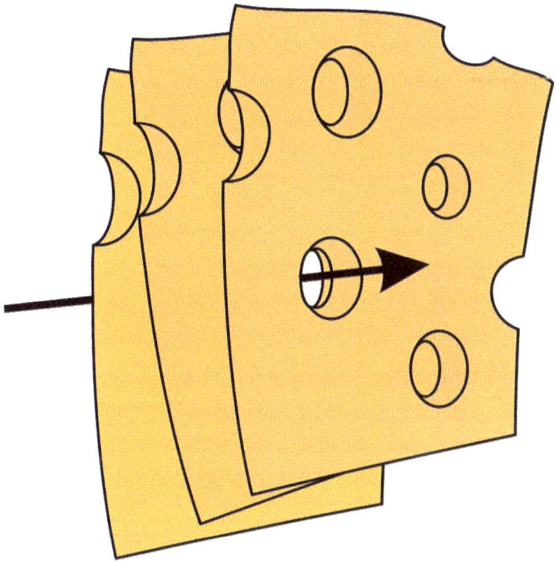

**Fig. 2.9** Reason's (1990) Swiss Cheese Model (a disaster is a sequence, which penetrates all layers. Layers are safety measures.)

by a single failure, but by a cascade of failures. Reason (1990) has introduced such a model consisting of different layers or barriers (Fig. 2.9). Sometimes it is called the Swiss Cheese Model, where a disaster only occurs, if all Cheese levels are penetrated by a sequence. The origin of the topic of safety culture is based on a large failure of safety culture causing the Chernobyl disaster (Ostrom et al. 1993), however it is now also widely applied in other fields such as Aviation Safety.

This approach is in compliance with the Integral Risk Management Approach for natural risks (Kienholz et al. 2004), Living Safety Analysis for technical risks or Risk Informed Decision processes for political risks.

## 2.4 Conclusions

Both, this chapter "Categorization of Safety and Risk" and the former chapter "What is Safety and does Optimal Safety exist?" show, that our understanding of safety and risk is still limited. This is mainly based on the fact, that both terms are extremely interdisciplinary and that they lack a common understanding. They are constructed based on life experience, educational background, cultural background and social conditions. This is even true for so-called mathematically-statistically approaches, since they are indeed based on an educational and scientific background. We have to face the fact, that there exists no such thing as an objective risk. Since the definition of complexity features a property such as "does include subjective judgment under all conditions" (Kastenberg 2005), obviously "risk" is an outcome of a complex decision process based on complex systems, including those persons or systems, which do the risk assessment. As mentioned in the beginning of this chapter, the risk assessment process is very much related to the complex system (human, society, computer) carrying out the risk estimation.

In other words, since risk estimation always includes information about the risk assessor, it can never be carried out objectively, since an objective statement should never include information about the decision maker (Todd and Gigerenzer 2003). This is not per definition a failure of the concept, since many items in our world do and must include subjectivity, but we have to be aware of this fact and we have to consider this, if doing a risk assessment. If one evaluates a risk, besides the rational risk assessment techniques including the determination of frequency, consequences and other subjective factors, we have to consider the following items:

- What are the consequences for us and for the risk assessor doing such an evaluation? Please be aware that herewith we do not mean the consequences of the risk itself, but the risk assessment process?
- Who is carrying out the risk estimation and does he/she gain any profit from the risk estimation?
- What is the causal relationship for the risk; is there any responsibility included?
- What are the alternatives in terms of actions related to the risk and can they really lower the risk?

It does not make sense to discuss alternatives at all, if the alternatives increase a risk, even doing nothing may increase the risk. Therefore, we have to talk about risk "forests" instead of single risks. We have to watch out for the largest trees and also for the fastest growing trees. Hence, the class of the tree (deciduous or needle tree) is not so much of importance, even if some trees may, in general, grow faster than other types of trees.

After all, returning to the conclusion of the chapter "What is Safety and does Optimal Safety exist?", the terms "safety", "risk", "quality of life" are based on social values. These values have to be defined by social debates and agreements. Even if it is every day business nowadays, that engineers or physicians decide about safety, they need a legal and social basis for their decisions since these decisions are neither perfect nor are they based on perfect information. By blaming certain professions, which is quite common now, we are actually increasing the risk since the general success of human societies is strongly related to the specialization of humans.

The promises of modern societies to provide indefinite low risks and indefinite strong support for every single member of the society are obviously impossible. Although we know that our societies are based on the trust of this promise, the acceptance of the limited capabilities of societies as a fact is the first step to decrease our own risks. One type of categorization of risk would be risks we can manage to a certain extend as an individual, risks we can manage as society and risks we can not manage. This categorization may rather sound non-scientific, but very useful. Even further, as the Swiss Cheese Model shows, things regularly go wrong - this is not a risk, this is daily life. The term risk is usually not used for such events. Therefore, risks are connected to unwanted outcomes reaching a magnitude not common. By introducing certain barriers of protection many events which were everyday life in former times are now risks in our understanding.

Another final remark: it should be a duty of any risk assessor to put risk into relation. Neither consider nor use a single risk, but rather use risk-comparisons or, even better, risk-benefit packages for comparisons. It is a very common fault, used every day in the news or other communication media to pick single risks and discuss their consequences. It is like saying a price of a good is to high without informing about the actual price. Therefore, always use bundles of risks.

## References


Covello, V. T., Peters, R. G., Peters, R. G., Wojtecki, J. G., & Hyde, R. C. (2001). Risk communication, the West Nile virus epidemic, and bioterrorism: Responding to the communication challenges posed by the intentional or unintentional release of a pathogen in an urban setting. *Journal of Urban Health: Bulletin of the New York Academy of Medicine, 78* (2), 382–391.

Dombrowski, W.R. (2006). Die letzten Meter – Kein Kommunikations- sondern ein Inter-aktionsproblem. 7. Forum und Gefahrentag. DKKV & GTZ, Eschborn.

Fischhoff, B., Lichtenstein, S., Slovic, P., Derby, S. L., & Keeney, R. L. (1981). *Acceptable risk.* Cambridge: Cambridge University Press.

Kaplan, S., & Garrick, B. J. (1981). On the quantitative definition of risk. *Risk Analysis, 1*(1), 11–27.

Kastenberg, W. E. (2005). Assessing and managing the security of complex systems: Shifting the RAMS paradigm. In *Proceedings of the 29th ESReDA seminar*, JRC-IPSC, Ispra, Italy, 25–26 October 2005 (pp. 111–126).

Karger, C. R. (1996). Wahrnehmung und Bewertung von "Umweltrisiken". Was können wir aus Naturkatastrophen lernen? *Arbeiten zur Risiko-Kommunikation.* Heft 57. Programm-gruppe Mensch, Umwelt, Technik, Forschungszentrum Jülich GmbH, Jülich: März.

Kienholz, H., Krummenacher, B., Kipfer, A., & Perret, S. (2004). Aspect of integral risk management in practice – considerations. *Österreichische Wasser- und Abfallwirtschaft, 56* (Heft 3–4), 43–50.

Klinke, A. & Renn, O. (1999). Prometheus unbound – challenges of risk evaluation, risk classification and risk management. Arbeitsbericht 153. Akademie für Technikfolgen-abschätzung in Baden-Württemberg, Stuttgart.

Metzner, A. (2002). *Die Tücken der Objekte – Über die Risiken der Gesellschaft und ihre Wirklichkeit.* Frankfurt : Campus Verlag GmbH.

Ostrom, L., Wilhelmsen, C., & Kaplan, B. (1993). Assessing safety culture: In general safety considerations. In G. T. Mays (Ed), Nuclear safety (Vol. 34, No. 2, April–June 1993, pp. 163–172).

Proske, D. (2009). *Catalogue of risks.* Heidelberg: Springer.

Reason, J. (1990). The contribution of latent human failures to the breakdown of complex systems. *Philosophical Transactions of the Royal Society of London. Series B, Biological Sciences, 327* (1241), 475–484.

Renn, O. (1992). Concepts of risk: A categorization. In S. Krimsky & D. Golding (Eds.), *Social theories of risk* (pp. 53–79). London: Praeger.

Renn, O. (1998). The role of risk perception for risk management. *Reliability Engineering and System Safety, 59,* 49–62.

Slovic, P., Fischhoff, B., & Lichtenstein, S. (1980). Facts and fears: Understanding perceived risk. In R. C. Schwing & W. A. Albers (Eds.), *Societal risk assessment: How safe is safe enough?* (pp. 181–214). New York: Plenum Press.
Starr, C. (1969). Social benefit versus technological risk. *Science, 165,* 1232–1238.
Todd, P. T., & Gigerenzer, G. (2003). Bounded rationality to the world. *Journal of Economic Psychology, 24,* 143–165.
Tversky, A., & Kahneman, D. (1974). Judgment under uncertainty: Heuristics and biases. *Science, 85,* 1124–1131.
Viscusi, W. (1995). Risk, regulation and responsibility: Principle for australian risk policy. Risk. Regulation and responsibility promoting reason in workplace and product safety regulation. In *Proceedings of a conference held by the institute of public affairs and the centre for applied economics*, Sydney, 13 July 1995. http://www.ipa.org.au/Conferences/viscusi.html.
WBGU. (1999). *Wissenschaftlicher Beirat der Bundesregierung Globale Umweltver-änderungen.* Welt im Wandel – Umwelt und Ethik. Sondergutachten, Metropolis-Verlag, Marburg.
Weichselgartner, J. (2002). Naturgefahren als soziale Konstruktion, Dissertation, Bonn, 2001.
Wiedemann, P. M. (1999). Risikokommunikation: Ansätze, Probleme und Verbesserungsmöglichkeiten. *Arbeiten zur Risikokommunikation*, Heft 70 Jülich, Februar 1999.

# Chapter 3
# Philosophical Perspectives on Safety and Risk

**Niels Gottschalk-Mazouz**

**Abstract** Humans care about themselves and their future. If nothing bad can happen, they are safe. If they know something for certain, they have certainty. But, as finite beings, humans do not enjoy much certainty. And in a dynamic world, they are rarely safe. "Risk" is one of the concepts that may help to come to terms with this. It is, however, notoriously unclear. That is why this text deals, in its first parts, with terminology. Seven basic uses of the term "risk" are distinguished, as well as some attributes like "negative", "hypothetical", "meta" or "systemic". The main objective of this text, however, is a discussion of rational and moral aspects of risk-taking, and of risk governance within what will be called "risk cultures". Culture, in this analysis, has three dimensions, a physical, an informational, and a social. Typically, only the adequate interplay of all the components in the three dimensions allows for a society, and an individual as its part, to adequately deal with risks. It is within—historically variable—risk cultures that risks are transformed. And it is a signature of modernity that risk cultures have become deliberate risk cultures.

**Keywords** Risk · Uncertainty · Subjective · Ethics · Culture · Philosophy

## 3.1 Introduction

Humans care about themselves and their future. If nothing bad can happen, they are safe. If they know something for certain, they have certainty. But, as finite beings, humans do not enjoy much certainty. And in a dynamic world, they are rarely safe. "Risk" is one of the concepts that may help to come to terms with this. It is, however, notoriously unclear. That is why this text deals, in its first parts, with terminology. The main objective of this text, however, is a discussion of rational

N. Gottschalk-Mazouz (✉)
Department of Philosophy, University of Bayreuth, Universitätsstraße 30, 95440 Bayreuth, Germany
e-mail: niels.gottschalk-mazouz@uni-bayreuth.de



M. Raue et al. (eds.), *Perceived Safety*, Risk Engineering,
https://doi.org/10.1007/978-3-030-11456-5_3

and moral aspects of risk-taking, and of risk governance within what will be called "risk cultures".

## 3.2 Risk: Some Basic Uses of the Term

The different uses of this concept can by systemised as follows (Gottschalk-Mazouz 2011; cf. Shrader-Frechette 1998): Risk means either (1) the *possibility* of an event, or (2) the *probability* of an event, or (3) the *value* of a possible or probable event, or (4) the *product* of probability and value of an event. Examples for (1) and (2) can be found in ordinary language talk like "there is a risk that something goes wrong" or "the risk that something goes wrong is pretty high", but also in Normative ethics or Bayesian decision theory, respectively. Examples of (3) and (4) can be found in Risk-benefit-analysis or Insurance mathematics, respectively. Moreover, in more generic ways risk has also been defined (5) as the result of some nonlinear function of probability and value of an event, or (6) as probability *cum* value of an event, or—maybe most comprehensively—(7) as the complete set of (relevant) event-probability-value triplets (Kaplan and Garrick 1981; Kaplan 1997). All these attributions, however, can be meant *objectively* or *subjectively*: For risk (1), objective risk means that bad things *can* happen, whereas subjective risk means that one *thinks* that bad things can happen, etc.

Of course, not only *bad* things can happen. But, when speaking of risk, very often we seem to be focussing exclusively on bad things. So, it is presupposed, then, that either nothing happens (if we are lucky), or that something bad happens (if we are unlucky). In other words, the event under scrutiny is a bad thing, a loss, a damage. With respect to risk (3) and (4), we then take the value of the event to be the extent of this damage. If the extent of damage is measured in some monetary value, the symbolic expression of risk (4) as $R = P \times E$ amounts to what is sometimes called the "insurance formula": Risk (R) equals probability of damage (P) times extent of damage (E). When we focus exclusively on bad things, we are framing risk in a negative way, because the value of the event can only be negative. The "greater" the risk the worse it is. That is why I call this the *negative* concept of risk.

But good things can happen as well: It is often said that we should not focus on what can go wrong but on the chances, or opportunities. This talk of chances or opportunities can be analysed in a completely symmetric way: With this it is presupposed that either nothing happens (if we are unlucky) or that something good happens (if we are lucky). This amounts to something that, if people were using the term 'risk' for it, should be called the *positive* concept of risk. But people rarely do.

Rather, what people do, is incorporate the chances into a broader concept of risk that is neither purely negative nor purely positive, and that I want to call *speculative* risk. With it, we allow for bad as well as good things to happen *within* the concept of risk. So what we presuppose then is that either a bad thing happens (if we are unlucky), or a good thing happens (if we are lucky), or nothing at all happens (which is taken to be neutral, i.e. is setting the baseline). We do so by allowing for negative or positive values of the events.

Now, the world is rarely such that events come alone. They come as clusters, or as consequences, i.e. as sets, which usually are of mixed value: some negative, some positive. Now, we have two equivalent ways of expressing that. Following the concept of negative risk, we can say that a given set of events contains risks and opportunities. If we know enough about it to compare them, we can say that it contains more risk than opportunity (or the other way around, or that they are equal). Following the concept of speculative risk, we would rather say that the set of events contains negative and positive risk. And if we know enough about it, that the *overall speculative* risk or the *risk balance* is negative (or positive, or zero). What we are saying in either of these two ways can be given the same precise meaning depending on what concept of risk is invoked. That there is more risk than opportunity, or that the risk balance is negative, would mean for example for risk (1) that more of the events are negative than positive.

To be able to form a risk balance, you have to be able to compare risks according to some apt metrics. The definitions of risk (1–4) point to respective natural metrics, however, which for risk (1) is a binary number, for risk (2) is a real number between 0 and 1 or any other probability measure, for risk (3) is a monetary value or any other cardinal (or, for that matter, at least ordinal) value, and for risk (4) is the product of probability and value, which is typically also a value, namely a cardinal value.

However, events do not form such clusters by nature, at least not in a sense that can be used in balancing risks: Causal chains (or causal networks) stretch out indefinitely in time and, in a connected world, get very complicated to trace even for relatively short timespans. So every question about risks has to be constrained according to some background relevance conditions and standards. If you ask for a possibility, as in risk (1), it has to be made clear what is taken to be constant (e.g. natural laws, social structures, individual behaviour etc.) and what is taken to be variable (e.g. environmental conditions). Moreover, of course, you need standards of demarcation that allow you to give a yes/no answer. Only then does it make sense to ask whether an event x is possible or not (e.g. that your house will suffer from flooding). If you ask for a probability, as in risk (2), further constraints have to be introduced: The standard way to understand probabilities is to see them as relative frequencies, i.e. ratios of numbers of events (e.g. the weeks where your house will suffer from flooding divided by the total number of weeks). Now we need a standard of division (e.g. the weeks). If the risk definition involves evaluations, as in risk (3) and (4), it is clear that an evaluative standard (as one of many possible standards) is presupposed as being adequate for evaluating these events (e.g. a monetary value of the damage due to flooding). All these standards might be controversial, which may be evident for evaluations (e.g. shall we monetarize when loss of lives is involved?), but is also the case for divisions (e.g. shall traffic accidents be counted per year or not better per 10.000 km?) or demarcations (e.g. from where on shall we speak of "flooding", or of a "traffic accident"?). This is virulent in all the risk comparisons that we frequently make, e.g. travelling by plane might be riskier than by car if compared by the hour, but is less risky if compared by distance travelled.

## 3.3 Risk: Objective, Subjective, and Perceived

Subjective risk is not necessarily the same as perceived risk. As the terms have been introduced, objective and subjective denote kinds of properties of events. "Objective" means that something is attributed with respect to the object, i.e. that it lies in the nature of the object whether it is correctly attributed, and this is what makes it true or false. "Subjective" means that now it lies in the nature of the subject, i.e. its beliefs, expectations or desires, whether it is correctly attributed, and this is what makes it true or false.

In the case of risk (1), objective risk would mean that something can happen. It is an objective question, independent of what we believe or want, whether this is the case or not. Subjective risk would mean that we expect that something can happen. In the case of risk (2), objective risk would mean either that we apply some concept of objective probability for single events or that we follow the concept of relative frequencies to come up with some ratio. Subjective risk would mean that we assign subjective probabilities or ratios, i.e. expectations that depend on what we believe to be the case or what will be the case. The concept of objective probabilities for single macroscopic events is metaphysically murky, though, and psychological research pointed out that frequency (ratio) formats are cognitively superior as well (Gigerenzer and Hoffrage 1995). In the case of risk (3), an objective risk would mean that the evaluation is objectively true, i.e. that a given event would, objectively, cost a certain number of lives, if "lives lost" is our evaluative standard. Subjective value would mean that we expect it to cost those lives. Finally, risk (4) might be purely objective, subjective or hybrid.

"Objective risk" does not mean that no subject's activity is involved in assigning it, however. To the contrary, because properties are not self-attributing, there would be no instance of any concept objective or subjective without a subject that performs attributions. It is just that the attributions are meant objectively, to be true or false in virtue of the things that there are. "Subjective risk", on the other hand, does not mean that the standards of attribution become subjective, but whether that which is attributed is a property, ultimately, of the subject and not the object. That there is subjective risk (3) e.g. does not mean that one person measures value in dollars and the other one in lives lost. But it means that we talk now about *expecting* something to cost some amount of money if it happens, or some number of lives. So even though evaluation needs an evaluative framework, which might be man-made, like money, it might nevertheless be an objective fact that some damage costs us some amount of money to repair it—whether we know it or not.

Because we, as human beings, have no direct access to the world as such, in its pure objectivity, the distinction between objective and subjective risk might seem purely academic. But it is not. First, objectivity can be made sense of as the asymptotic limit of intersubjectivity, for example in the pragmatist sense of an ultimate opinion that a community of investigators comes to form over time (Peirce and Charles 1909). Second, subjectivity can be said to be a matter of degree: Combining both points, we can say that some risks may be more objective or more subjective in character depending on the level of intersubjective conformation of the supporting beliefs or values.

"Perceived risk" is ambivalent. It can either be understood as a demarcating label, such that there are perceived risks and other, maybe: real, or hidden, or whatever risks. But perceived risk can also be understood as an explanatory label, such that every risk is a percept, maybe even a perception of a perception (or of perceptions). In any case, speaking of perception is presupposing that there is something to perceive, and that we might perceive it (and perceive it correctly or adequately) or not. Understood as such, it draws from the same distinction as objective and subjective risk. There are three options to further specify what is perceived: It may be either objective risk, or any other objective entity, or a subjective entity. The latter two options are compatible with both the explanatory and the demarcating understanding of "perceived risk". The first option is compatible only with the demarcating understanding of "perceived risk", though.

As for "perceived safety", this expression usually does not mean that adverse events are believed to be impossible. If German politicians say, "Our nuclear power plants are safe!" or "Our dams are safe!" it is clear that they cannot mean that literally; they can only mean that either the probability that things go wrong is close to zero, or that the possible damage is close to zero, or both. And we may believe this or not. But they cannot mean that one (or both) are really zero. Perceived safety, in this sense, would always be a misperception. A better alternative would be to understand safety as practical safety: Something is practically safe when anything that might go wrong is either so improbable or causing so little damage that it makes no difference to our choices. Some people might live in perceived safety in this sense; they think that they need not worry. And while this is not always a misperception, it is well known that it sometimes is.

## 3.4 Risk and Uncertainty, Hypothetical and Meta-Risks

Depending on the chosen definition of risk, it might make sense to differentiate between risk and uncertainty. Certainty and uncertainty are subjective concepts, so uncertainty and *objective* risk are not of the same kind. It makes more sense to compare subjective risk and uncertainty. We can be uncertain about many things, but for our purposes, it might be useful to say that "uncertainty" is more radical than risk. Sometimes uncertainty is used to describe anything that is less determinate than risk (4). So, if we know that something might happen, but do not know its probability or value, we have uncertainty and not risk (4). But what we have, then, can be easily spelled out by taking the other definitions of risk (1–3) into account. Namely, if we know that something might happen, but not the probability or value, then this is the situation of risk (1). If we know its probability or value, but not both, this is risk (2) or (3) respectively.

So, the demarcation between risk and uncertainty is relative to the involved definition of risk. For any risk (n), what is less certain turns out to be a risk (m) with m < n. Only for something that is less certain than risk (1), do we need a new concept. This could mean that we do not expect that something bad or good might happen at all, or that something bad or good of a certain kind might happen (so that we can, at least qualitatively, distinguish it from other possibilities, i.e. in its essence, but not in terms of probability or value). For the latter, I would thus suggest to use the term "essential uncertainty", and for the former the term "complete uncertainty" or "ignorance". Please note that ignorance does not mean that we do expect that nothing at all happens, it means that we expect something to happen for sure (business as usual, so to say) and thus we do not expect that something *might* happen (or not happen), or, in other words, we do not expect any surprises at all. Essential uncertainty means that, after something unexpected has happened, we would say: "I did not expect that it might go wrong in *that* way". Ignorance, however, means that we would have to say: "I did not expect that anything *at all* might go wrong here".

These extents of uncertainty have to be distinguished from orders of uncertainty. First-order uncertainty about something, say a damage extent, means that we do not know the exact value that quantifies it. Sometimes, we can quantify this first-order uncertainty by using tools and concepts from statistics, e.g. by using error bars. Second-order uncertainty would mean, however, that we do not know how exact we know (or not know) the value in question.

A further distinction, that is cross-cutting to those explained so far, is that between real, hypothetical and meta-risks (cf. Hubig and Christoph 1993: 75). A risk is real, in the sense of this distinction, if the domains of definition are well-known and the events under scrutiny uncontroversially fall in these domains. A risk is hypothetical, however, if this is not the case. The events under scrutiny are only theoretically described (and we lack the experience that would reassure us that we got the theory right), or these events are only qualitatively observed such that quantities have to be estimated etc.—and hence we cannot be sure that the events will be confined to the domains of definition. That a risk is hypothetical, in the sense just introduced, does not mean that we should take it less seriously or that we do not really know what to expect. In the case of meta-risks, however, the event under scrutiny has the potential to change the domains of definition for other risk evaluations, such that many other risk assessments will have to be revised in a typically unclear way. It should be clear that these risks must not be neglected. An example for a hypothetical risk would be the release of genetically modified organisms into the wild when we know about their behaviour only from model calculations, theoretical estimates and analogies to other organisms' behaviour in the wild. Examples of meta-risks would be the triggering of massive climate change, the creation of a hybrid species between animal and human, or of some singularity-like self-enhancing artificial intelligence.

## 3.5 Risk and Choice

It has been suggested to distinguish between risk (Risiko) and danger (Gefahr): While danger, it is said, concerns events that might happen independent of what we do, to speak of something as a risk means that our own decisions, our choices, are involved (cf. Evers et al. 1987). The neat example, following Luhmann and Niklas 1993, is that before there were umbrellas, given that I had to go out, there had been a certain danger to get wet. With an umbrella at hand, this danger turns into a risk. Now I can lower the risk of getting wet, by taking the umbrella with me, but I run other risks with that, e.g. the risk of losing the umbrella. In a social setting, on the other hand, consequences of a decision may, for those that decide, appear as a risk, what for those affected appears as a danger (Luhmann and Niklas 1991).

While the umbrella story nicely illustrates one of the many repercussions and side-effects of technology, or more general, of the use of means, and the risk/danger distinction may be helpful for social theory, I do not think that ordinary language use warrants the use of the labels "danger" and "risk" for it. After all, we say that the risk of a fatal hit of the earth by a meteor is so and so (we mean: the probability, and may presuppose that "fatal", i.e. the damage under scrutiny, is the extinction of mankind). Now, one can say that we should not use the word "risk" here, but then I do not think that "danger" would be a viable alternative. Yes, we can say that this danger is greater than that danger. But nobody would say: "The danger of … is 0.01 per year" or "The danger is 100 EURO") or the like. Rather, the best alternative to risk jargon would be to say "The probability of … is 0.01 per year" or "The possible damage is 100 EURO". So, danger talk cannot fully substitute risk talk *regardless* of whether the outcome is dependent on our choices or not. Thus, I do not see that danger and risk are alternative concepts, but rather that for some negative risk, in the sense of risk (1), we also use the term "danger", as well as for some undifferentiated risk (2–5)-talk. Hence, I will stick to the risk-jargon.

Independent of such labelling problems, the connection between risk and choice has been thoroughly discussed in economics and ethics in terms of rationality. By definition, one should try to avoid bad events, so vis-à-vis *negative* risks, the rational behaviour is aversion: After all, who would prefer a 50/50 chance of losing 1 Euro (i.e. a 50% chance of losing the Euro and a 50% chance of keeping it) over a 100% chance of not losing or gaining any money? When economists speak of "risk aversion" (cf. Daniel and Tversky 1979), however, they refer to what I called *speculative* risk. Then, aversion means to shy away from uncertain outcomes just because of their uncertainty. So if you are risk averse in the economists' sense, you would prefer a 100% chance of not losing or gaining any money over a 50/50 chance of losing *or gaining* 1 Euro. If you are in a situation where you cannot afford to lose 1 Euro (e.g. because you need it to buy your next meal) but one more Euro would not make much of a difference to you, it would be rational to be risk averse. Whereas, when one Euro less would not harm you but 1 Euro more would make a big difference (e.g. with just this Euro more in your pocket, you could buy what you always wanted), it would be rational to be risk seeking.

This rational risk seeking (or risk aversion) behaviour is not to be confused with the pleasure (or disgust) that some people might find in gambling. Because if you love safety for its own sake and hate gambling, you might prefer not to gamble at all over a gamble that promises you only wins. Or, if you really like gambling as such, you would be willing to gamble in situations where, say, you have a 49/51 chance of winning or losing a Euro, or where every outcome makes you lose something (but the grief from losing some money is outweighed by the intrinsic pleasure from gambling). So, if one does not confine rationality to making or maximising profits (in some narrow sense, e.g. of winning of losing money), this latter behaviour is also not irrational. It is then guided by *intrinsic* risk aversion.

## 3.6 Rational Choice Under Risk and Uncertainty

In situations that can suitably be described as risk (4), i.e. where the values and probabilities of the options are known, and the according demarcations, divisions and evaluations are uncontroversially adequate and inclusive, i.e. that all relevant short- and long-term effects are accounted for, all ex-ante as well as ex-post costs, e.g. for precautionary measures or damage repair, are included in the evaluation, and also all subjective costs due to persistent fear or dread etc. are included in the evaluation—in such an ideal case, risk (4) can be interpreted as the expected value of the event. Rational choice theory then tells us that it is rational to go with positive expected values, or in the case of multiple options, that It would then be most rational to always carry out what promises the highest expected value (sometimes dubbed the "Bernoulli-Principle"), cf. Mongin and Philippe (1997).

Real-world risk calculations, however, are typically considerably more narrow in scope. In them, one looks at only certain types of consequences (e.g. life, or property), and at only the negative consequences (loss of life, or property), and at aggregates along quite limited dimensions (typically: by counting loss of life, or by summing up monetary values of property loss, or by combining both, and possibly more such as disease incidents, with exchange rates based on willingness-to-pay or welfare loss, cf. World Bank (2016), 47ff.). With respect to expected value, these calculations are necessarily incomplete, for above all, they do not include chances: they are only expressions of negative risk, not of speculative risk, and moreover only of certain aspects of it. Because of this, the rejection of different risks and the acceptance of equally high risks *in any such calculatory framework* can usually not be regarded as irrational, but may well be a very rational attitude of encompassing the aspects that are neglected by the framework.

Moreover, every interesting real-world calculation of a risk (4) will fall more or less short even of the narrow-scope ideal. Typically, it has to deal with imprecise numbers, estimates and guesses. Most challenging are events that are rare or that occur within complex systems (more on that below), are because the proper basis of experience is missing or the causal analysis does not work, respectively. There are some strategies of dealing with such uncertainties that are worth explaining in more detail. While the discussion, and some of the titles, are centuries old (cf. TNCE 2003), the framing given here is modern and uses the above distinctions: *Probabilism* would attempt to assign probabilities and values despite these uncertainties, i.e. determine or construct risk (7), and then decide for one of the options. It comprises a variety of substrategies that differ in how to do the assignment. One such substrategy would be *probabiliorism*, which consists of identifying the most probable consequences and act as if they were certain (and not only quite probable). Indicators of such a strategy are justifications that include phrases like "it is virtually certain that…" or "it is beyond reasonable doubt that…". Another substrategy would be "Bernoulli-Maximisation": here we assign equal probabilities to all unknown consequences, deal similarly with damage extents, and follow the Bernoulli-Principle (i.e. choose the option with the highest expected value). *Tutiorism* is an alternative to probabilism. Here we do not assign any probabilities at all, but look only at the values of the consequences, i.e. the extent of damage. We then try to avoid the option (or those options) that would bring maximal damage or would choose those options that would bring the maximal relief from suffering. The maxim recommended by the German philosopher Hans Jonas to give priority to the worst outlook ("Vorrang der ungünstigsten Prognose", Jonas 1979) can be seen as an example of this strategy. To judge a measure only with regards of how it harms or benefits those that are worst off (as the Rawlsian "difference principle" suggests, cf. Rawls 1999), or most vulnerable, in a society would also be a substrategy of tutiorism.

As soon as hypothetical risks or meta-risks are involved, all risk quantifications including risk (7) are becoming more and more insignificant and inadequate. In these cases, qualitative rules or scenario-based methods might provide better guidance. Ideally, we can then characterize the risk and identify best ways of dealing with them vis-à-vis the full spectrum of uncertainties. One proposal of a set of qualitative rules comes from the WBGU (1999), which takes into account second- and third-order uncertainties, and uses criteria such as ubiquity, persistence, irreversibility, delayed effect and mobilisation potential for the characterisation of risks and the recommendation of adequate strategies. Finally, scenarios can help to structure the possibility space and, thus, help to identify or construct meaningful options that can then be qualitatively or (sometimes) quantitatively compared. Especially for meta-risks, this may sometimes be the only viable option.

## 3.7 Risk Acceptance and Risk Acceptability

Risk acceptance is a descriptive term that refers to observed behaviour (risks people take) or judgements (risks people are willing to take). In social settings, however, my choices affect others and I am affected by theirs—and I know that. This creates some complexity in understanding risk behaviour/judgements. Very often, risk decisions are to some degree *allocentric*, "other-directed", i.e. those that decide about an action are not identical with those that are affected by the possible costs or benefits of this action. The risk-taker is not identical to the risk-sufferer, or so one may say. And, to a certain degree, each side takes into account the perspective of the other. In attributing risk perception, the question would be in regard to whose perceptions are involved (or, more generally, to which subject the subjective components shall be related). Another way in which a risk may be "social" is when the possible benefits and the possible harms do not affect the same persons. The risk may be *asymmetric*, i.e. individual risk-balances may be diverging. And again we have to ask to which subjects any subjective elements shall relate.

Every actor knows that he or she sometimes decides for others (e.g. children) the risks to take, and that those who may benefit are sometimes not those who may suffer. Thus, in our judgements about the risks we want to accept, as well as in our behaviour in dealing with risks, we have taken the social dimensions of risk somehow into account, i.e. we have taken some stance towards the other. Empirical research in psychology and sociology may help to describe risk behaviour and judgements, but it not easy to interpret these findings because of the complexity of decisions under risk and uncertainty.

Acceptability of risk, on the other hand, is a normative term. It means possible acceptance, not in a prognostic sense but in the sense of meeting considered judgements about legitimate acceptance. These judgements are normative judgements because they involve either prudential evaluations (about what is good for me, or for us as a group) or moral evaluations (about what is good, or even equally good, for all of us). Prudential evaluations are, in their general form, linked to conceptions of a good life—and it is in the light of these conceptions that strategies of dealing with risk and uncertainty (including the strategies explained above) are chosen. Psychology and sociology can help to understand these links; philosophy should also be able to contribute, but it is traditionally more occupied with moral evaluations. These evaluations are mainly discussed within the frameworks of the grand moral theories, utilitarianism, deontology and contractualism. Utilitarianism typically sees the maximization of the sum over everybody's expected value as the right thing to do. It is then sensitive only to the sum, not to the distribution (between individuals) of such a value, and it may well be that risk decisions make some people worse off as long this is compensated by making some other people better off. This approach connects to economic theory. Deontology, however, operates with indexical rights of individuals that do not allow for such calculus. The infringement of rights has to be justified by the exertion of some other, equally high or higher-indexed right, and that there are probabilities involved, does not change this principle (Nida-Rümelin and Julian 1996). So, e.g. if the right to live is ranked higher that the right to property, it is justified to impose risks of property loss on some people if this increases the chances of other people to survive. While in some constellations, it is still allowed to come to risk decisions that impose risks on some to the potential or immediate benefit of some others, equity considerations now require that those that are being made potentially worse off have to be compensated by those that benefit, or at least those that are being made actually worse off (if things go wrong and damage occurs). This approach connects to legal theory. Contractualism relies not on calculus or on antecedent rights to determine what is allowed or required to do, but on the agreement of those that are affected. The individuals shall be put in a position where they can make an informed decision as to which risks they are willing to take (vis-à-vis certain compensations etc.), and in cases when an individual decision is not possible, individuals should be in control of the institutions that govern the risk that do affect them (Renn et al. 2007). This approach connects to the theory of democracy (Shrader-Frechette and Kristin 1991), and has led to a variety of models of citizen participation in shaping collective risk decisions (Gottschalk et al. 1997). It can be grounded in "discourse ethics" (ibid.; Skorupinski et al. 2000) that contains deontological and, if not utilitarian, then, at least, consequentialist elements.

## 3.8 Risk Governance, Systemic Risks and the Precautionary Principle

"Risk governance" has more recently been introduced as a label for the wide-scope attempt to deal with risks in a rational and equitable way (Renn 2008) following the lead of the US National Research Council (National Research Council NRC 1996). It consists of four consecutive phases: pre-assessment (framing, early warning, screening), appraisal (estimation of hazards and exposures, assessment of risks), characterization and evaluation (see the last sections of this text), management (selection of options). Communication, in this framework, is a cross-cutting task in all four phases.

Risk governance in modern societies has become quite challenging: modern societies are organized as a network of very sophisticated institutions, are relying in many ways on advanced technology, and are interfering with the natural environment long-term. Risk decisions often have consequences that run through all three of these social, technological and environmental systems, which are coupled in many ways. Accordingly, concepts of "systemic risk" have been introduced. That risks occur (and have to be governed) in these systemic structures has been pointed out by the International Risk Governance Council: "Systemic risks are at the crossroads between natural events (partially altered and amplified by human action such as the emission of greenhouse gases), economic, social and technological developments and policy-driven actions, both at the domestic and the international level." (International Risk Governance Council IRGC 2005, 81). On the other hand, some of these systems are itself essential for the fulfilment of basic human needs. The OECD has pointed out this aspect: "A systemic risk [...] is one that affects the systems on which society depends—health, transport, environment, telecommunications, etc." (OECD 2003, 30). So, "systemic risk" may mean that risks occur due to systems (they are creating risks) as in the IRGC definition, or that systems are in danger due to risks (they are affected by risks) as in the OECD definition. In fact, both may be the case, due to the multitude of systems involved.

As the major challenges of modern societies in governing risks, it has been suggested to distinguish complexity, uncertainty and ambiguity (Klinke and Renn 2002; International Risk Governance Council IRGC 2005). Complexity means that system behaviour involves feedbacks and triggers effects that makes it hard to predict medium- and long-term behaviour, and to identify causal relations, i.e. describe them in terms of if-then-statements (which would be needed to discuss interventions). Uncertainty, used in contrast to complexity, points out not the objective complexities of these systems, but the subjective ignorance in determining the exact states of these systems, and their change with time. The interplay of complexity and uncertainty is well known from physics, where simple models already show deterministic chaos: they render very complex dynamic results that change qualitatively due to only minor changes in the control parameters, such that any uncertainties about these parameters make these systems practically unpredictable.

Finally, ambiguity should be an umbrella term for all evaluative and normative degrees of freedom. Modern societies are pluralistic, so common values for collective decisions are hard to find. On top of that, risk-specific ambiguities occur, i.e. due to diverging levels of risk-adversity. On the level of moral theory, a similar pluralism occurs (that of utilitarianism, deontology and contractualism, e.g.). A major problem is the time horizon of many risk decisions (among the most prominent here, are those concerning nuclear energy). It is disputed whether future events shall be discounted and, if so, at what rate. Moreover, evaluations may change with economic and scientific progress. What we see as worthless today may become a precious resource in the future, and what we see as harmless may be considered dangerous. Finally, it would be parochial to assume that values and moral ideals will not change with time, but we have no idea how they will change. However, we cannot simply take the preferences of future generations to be identical to our own preferences—we should allow them to form their own preferences. This makes it very hard to meaningfully evaluate risks - even if all natural consequences were known.

Suitable reactions to this threefold challenge include openness and explicitness in risk-governance. For risk identification and characterisation, this means that the underlying uncertainties are not ignored (e.g. when some parameter is not known exactly, a plausible distribution of parameter values should be used and not just the best estimate) and they are made explicit (e.g. by depicting error bars for any quantity, or by adding a verbal commentary that points out known unknowns, simplifications, second-order uncertainties etc.). Underlying ambiguities should be handled in a similar way by making divergent evaluations explicit and by laying open diverging or converging normative standards due to competing moral theories, etc. Openness and explicitness are essential components to the ideals of a "discursive" risk governance (Renn et al. 2007, 234).

Moreover, it is often said that we should follow a precautionary principle. This principle can be interpreted in different ways. In its most simple form, it says that we should try to avoid taking risks when the potential damage is catastrophic. These risks are, in the sense outlined above, meta-risks and the principle recommends a tutiorist strategy. Such catastrophes threaten our agency, i.e. narrow down our future options and our abilities to identify and evaluate them—we would merely be able to react to external circumstances in a daily struggle for survival and can no longer properly plan ahead (or, for that matter, assess risks) at all. In a more complex form, the precautionary principle says that we should also try to avoid taking severe risks, i.e. those with high damage potential, when the potential gains are in the same order of magnitude. This can be seen as an answer to complexity, uncertainty and ambiguity, where we should be careful if we do not really know what we are doing. A reason for this may be that what we see as potential gains (e.g. what we find pleasure in) are usually more variable than what we see as potential losses (e.g. life and limb, avoiding pain). And while we do not know what future generations will find enjoyable, we know well enough what will make them suffer. Both versions of the precautionary principle can be justified within any of the discussed moral theories, not only within deontology, although different reasons would have to be given and different concrete advice of levels of extra care would result.

Sometimes the term "precautionary principle" is used in a less abstract sense, however, namely as a label for regulative strategies that allow any innovation, only if proven to be (relatively) harmless. This is combined with a tort law that allows only for minor compensation claims in the case that (by definition) unforeseen risks materialize. On the contrary, a "risk principle" would allow any innovation if not proven to be (relatively) harmful. This is combined with hefty compensation claims. Both models allow one to deal with an open future and set checks and balances against irresponsible or egoistic risk behaviour. The former model is more in line with deontology and the latter with utilitarianism, and it is sometimes said that Germany (or the EU) would follow the former and the US the latter principle. What seems rather dangerous, however, is to combine the liberal admission policy of the risk principle with restrictive tort law, i.e. only minor compensation claims. On the other hand, restrictive admission plus high compensation claims may mean forfeiting too many opportunities. Thus it might be difficult to harmonize regulative frameworks across cultural borders.

## 3.9 Risk Cultures

Ultimately, this is what any society and any of its individuals has to acknowledge: that, nolens volens, something has been established that I would like to call, in a broad sense, a "risk culture". Culture, in this analysis, has three dimensions, a physical, an informational, and a social. In dealing with risks such as flooding (cf. Gottschalk-Mazouz and Niels 2008), the physical dimension includes dams, water buffers etc., the informational dimension includes public broadcasting and education, and the social dimension includes civil protection organisations, insurances, donations etc. Typically, only the adequate interplay of all the components in the three dimensions allows for a society, and an individual as its part, to adequately deal with risks. But for every culture, it is not only the static aspect (what the respective risk culture is like at the moment) but also the dynamic aspect (how it develops or is being developed) that deserves attention when dealing with risks: because risks change over time, new risks emerge and old ones disappear or become less relevant, and because social, technological and environmental conditions change as well, risk cultures are (or should bc) in continuous transformation.

This "risk culture" is not the same as the famous "risk society" hypothesis that Ulrich Beck has taken our modern societies to be. He has suggested (Beck 1992, 21) that risks are a "systematic way of dealing with hazards and insecurities induced and introduced by modernisation itself" and that we are thus on a way to a "New Modernity". That risks are induced and introduced (but also mitigated and absorbed) by modernisation may well be. But ever since man has been keeping stock, has been fabricating its own tools, has been communicating about adverse events and has been organised such as to collectively deal with them, there have been risk cultures. And in those cultures, risks have been transformed. Naturally, people would not have used these terms. And maybe they would have described what we take to be a risk as something else, some divine trial maybe, or something that could not have been altered (and has been predetermined) anyway. It is true that to develop *deliberate* risk cultures, modern concepts of human agency and choice, and modern concepts of nature and society are required. But as soon as these concepts were at hand, people have also been used the concept of risk (and, in fact, also the term "risk"; cf. Rammstedt and Ottheim 1992) to develop such deliberate risk cultures.

## References

Beck, U. (1992). *The risk society. Toward a new modernity*. London. (Orig.: Risikogesellschaft. Auf dem Weg in eine andere Moderne. Frankfurt/M. 1986).

Daniel, K., & Tversky, A. (1979). Prospect theory: An analysis of decision under risk. *Econometrics, 47,* 263–291.

Evers, A., & Nowotny, H. (1987). *Über den Umgang mit Unsicherheit: Die Entdeckung der Gestaltbarkeit von Gesellschaft*. Frankfurt/M.

Gigerenzer, G., & Hoffrage, U. (1995). How to improve Bayesian reasoning without instruction: frequency formats. *Psychological Review, 102*(4), 684–704.

Gottschalk-Mazouz, N. (2011). Risiko. In M. Düwell, C. Hübenthal, & M. H. Werner (Eds.), *Handbuch ethik* (3rd ed., pp. 502–508). Stuttgart.

Gottschalk-Mazouz, N. (2008). Risikokulturen. In J. Köngeter (Ed.), *Sicherheit und Risiko wasserbaulicher Anlagen* (pp. A1–A22). Aachen.

Gottschalk, N., & Elstner, M. (1997). Politik und Technik: Überlegungen zu einer innovativen Technikgestaltung in partizipativen Verfahren. In M. Elstner (Ed.), *Gentechnik, ethik und gesellschaft* (pp. 143–179). Heidelberg.

Hubig, C. (1993). *Technik und Wissenschaftsethik: ein Leitfaden*. Berlin/New York.

International Risk Governance Council IRGC. (2005). Risk Governance: Towards an Integrative Approach. White Paper#1. Geneve.

Jonas, H. (1979). *Das prinzip verantwortung*. Frankfurt/M: Versuch einer Ethik für die technologische Zivilisation.

Kaplan, S. (1997). The words of risk analysis. *Risk Analysis, 17*(4), 407–417.

Kaplan, S., & Garrick, B. J. (1981). On the quantitative definition of risk. *Risk Analysis, 1*(1), 11–27.

Klinke, A., & Renn, O. (2002). A new approach to risk evaluation and management: risk-based, precaution-based and discourse-based management. *Risk Analysis, 22*(6), 1071–1094.

Luhmann, N. (1991). *Soziologie des Risikos.* Berlin, New York.

Luhmann, N. (1993). Die Moral des Risikos und das Risiko der Moral. In G. Bechmann (Ed.), *Risiko und gesellschaft. grundlagen und ergebnisse interdisziplinärer risikoforschung* (pp. 327–338). Opladen.

Mongin, P. (1997). Expected utility theory. In J. Davis, W. Hands, & U. Mäki (Eds.), *Handbook of economic methodology* (pp. 342–350). London.

Nida-Rümelin, J. (1996). Ethik des Risikos. In J. Nida-Rümelin (Ed.), *Angewandte ethik* (pp. 806–831). Stuttgart.

National Research Council NRC. (1996). *Understanding risk.* Washington DC: Informing Decisions in a Democratic Society.

OECD. (2003). Emerging systemic risks in the 21st century: an agenda for action. Paris.

Peirce, C. S. (1909). Paragraph 184. In A. Burks (Ed.), *The collected papers* (Vol. 8). Cambridge, MA (1958).

Rammstedt, O. (1992). Risiko. In J. Ritter, & K. Gründer (Eds.), *Historisches Wörterbuch der Philosophie* (Vol. 8, pp. 1045–1051). Basel, Stuttgart.

Rawls, J. (1999). *A Theory of Justice (Revised Edition).* Oxford.

Renn, O., Schweizer, P.-J., Dreyer, M., & Klinke, A. (2007). Risiko. München.

Renn, O. (2008). *Risk governance. Coping with uncertainty in a complex world.* London: Routledge.

Shrader-Frechette, K. (1998). Risk. In E. Craig (Ed.), *Routledge encyclopedia of philosophy.* London, New York.

Shrader-Frechette, K. (1991). *Risk and rationality.* Berkeley, CA.

Skorupinski, B., & Ott, K. (2000). *Technikfolgenabschätzung und Ethik.* Zürich.

TNCE. (2003). In T. Carson, et al. (Eds.), *The new catholic encyclopedia* (2nd ed.), Washington, DC.

WBGU. (1999). Wissenschaftlicher Beirat der Bundesregierung Globale Umweltveränderungen: Welt im Wandel. Strategien zur Bewältigung globaler Umweltrisiken. Jahresgutachten 1998. Berlin, Heidelberg, New York.

World Bank. (2016). *The cost of air pollution: Strengthening the economic case for action.* Washington, DC: The World Bank.

# Chapter 4
# Psychological Perspectives on Perceived Safety: Social Factors of Feeling Safe

**Eric Eller and Dieter Frey**

**Abstract** What makes people feel safe? How do people conclude whether a certain situation, choice, or behavior is safe or not? In the present chapter we take the view that social factors influence perceived safety. We discuss the social determinants of perceived safety both as a general subjective state and as a safety-related estimation or judgment. From this perspective, we first discuss what humans *need* to feel safe. We present psychological insights on basic human needs and argue that the fulfillment of those needs is a general condition for the state of perceived safety. Second, we discuss how social factors (i.e. what others do and say and how one relates to these others) influence safety judgments and decisions. We illustrate how individuals adapt their judgments and behaviors to group norms and discuss why group discussions can lead to extreme judgments and decisions. We aim to complement the existing literature on perceived safety by highlighting the importance of social factors of safety perception.

**Keywords** Perceived safety · Social influences · Human needs · Need to belong · Social identity · Group conformity · Group polarization · Safety judgment

## 4.1 Introduction

Do you feel safe? Before reading on, please spend a moment to answer this question for yourself. You are probably able to quickly make a rough statement on whether you generally feel safe or not. However, safety is a complex, multi-layered construct, and the question can be answered in multiple ways. Safety refers to very different areas of human life, such as one's current health status, experienced

E. Eller (✉) · D. Frey
Department of Psychology, Ludwig-Maximilian-University Munich, Leopoldstr. 13, 80802 Munich, Germany
e-mail: eric.eller@psy.lmu.de

D. Frey
e-mail: dieter.frey@psy.lmu.de



M. Raue et al. (eds.), *Perceived Safety*, Risk Engineering,
https://doi.org/10.1007/978-3-030-11456-5_4

exposure to crime, financial situation, and social relationships. We argue (and this is a main emphasis of the present chapter) that social relationships are of importance for a person's well-being in general and safety perception in particular. For example, students could feel unsafe at school or university because they fear not being accepted by their peers or because they are afraid of not meeting the given performance expectations. Likewise, interpersonal conflicts within one's family or at work might lead to the experience of fear and uncertainty. The list of such social factors that can make us feel safe (or not) goes on. In the first section of the present chapter, we aim at going into detail about what humans *need* to feel safe, with a particular emphasis on interpersonal aspects.

In the second section, we discuss how safety-related judgments and decisions can be shaped by social factors. When people face safety-related questions such as whether certain sports, foods, medical treatments, journeys, or investments are safe, they are typically not in isolation. Instead, people tend to make such judgments and decisions in social contexts, as part of groups such as their families, friends, colleagues, or simply the people who surround them by chance in the train, at the airport, or in the supermarket. People often adapt their judgments and decisions to what is suggested or seen as normal by their surrounding groups. In the second section of this chapter we discuss a number of insights from social psychology that help us understand how social contexts can affect safety perception.

## 4.2 What Do We *Need* to Feel Safe?

Humans are loss-averse and thereby motivated to make decisions in such a way that losses are avoided or minimized. Per prospect theory (Kahneman and Tversky 1979; Tversky and Kahneman 1973), human decision-making is to a considerable extent based on the subjective perception of losses and gains. Humans are more sensitive to losses than to gains and thereby loss-averse. Safety perception relates to the motivation to avoid losses because it generally describes a state of protection from harm that is presently experienced as well as expected for the future. The etymology of the term safety (Old French: sauveté, Latin: salvus) refers to a condition of not being in danger, of being unharmed (see Oxford Dictionaries 2015). In order to understand the conditions under which humans feel safe and unsafe, we therefore need to understand what really endangers and harms people—or, expressed positively, we need to know and understand the most fundamental human needs and how humans respond to the satisfaction and deprivation of these needs. We need to know what humans *need* to be able to feel safe. Thereby, we suggest broad concept of perceived safety as a state in which a person's most important needs are satisfied and it is expected that this state will remain stable.

The term *need* is well-established in psychological literature. As per Kurt Lewin's field theory, needs release energy, increase tension and thereby motivate a person to behave in a certain way (Lewin and De Rivera 1976). If a need is unfulfilled, one perceives actual harm (Baumeister 2012). "Not getting something you need means more and is worse than not getting something you merely want" (Baumeister 2012, p. 124). This specification highlights the immediate relation between basic human needs and safety, which we have introduced as a condition of being unharmed. Advances in psychological research during the last decades, especially within social psychology, have created a considerable knowledge base on what humans most fundamentally need. Aiming to widen the understanding of the conditions under which people generally feel safe, we first give an overview about various basic human needs. After that, we take a closer look at need to belong theory, which is particularly important concerning the perception of (social) safety.

### *4.2.1 Overview of Basic Human Needs*

Soon after Maslow (1943) proposed his hierarchy of needs, the theory became widely known and was later described as one of the most popular theories in the literature of management and organizational behavior (Wahba and Bridwell 1976). The core of Maslow's theory consists of five need categories ranked in a hierarchical order. The needs pyramid illustrated in Fig. 4.1 became the symbol of recognition for the theory—although Maslow himself never used that illustration. The wide acceptance and popularity of Maslow's theory is especially interesting considering that there is to date sparse empirical evidence supporting the theory. There is especially poor evidence for the suggested hierarchy (i.e., that people seek

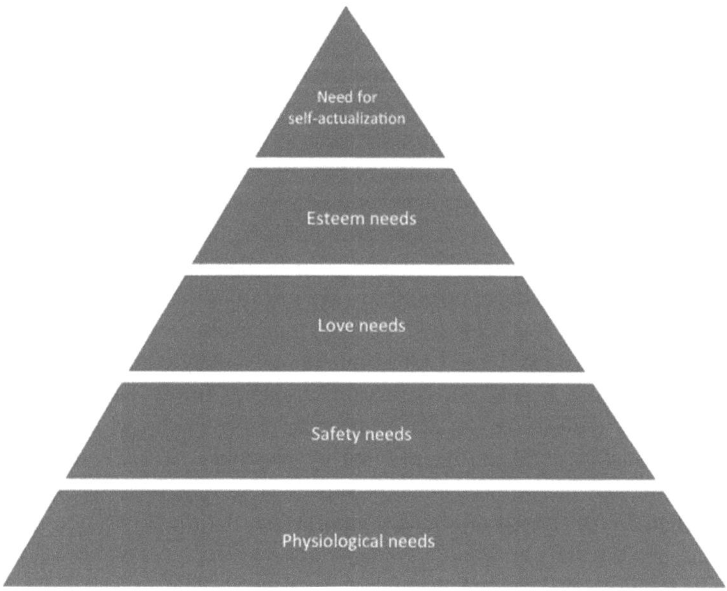


**Fig. 4.1** Maslow's hierarchy of needs (Needs categories as introduced by Maslow 1943)

need satisfaction in a specific order; Wahba and Bridwell 1976). However, Maslow's hierarchy of needs illustrates the numerousness and diversity of human needs and provides a broad overview of a number of them. Thus, the theory can be a good starting point to understand what really matters to humans.

The following five needs categories form Maslow's hierarchy of needs (1943):

1. *Physiological needs.* Maslow described physiological needs, including hunger, thirst, and fatigue, as the most pre-potent needs category. "For the man who is extremely and dangerously hungry, no other interests exist but food. He dreams food, he remembers food, he thinks about food, he emotes only about food, he perceives only food and he wants only food" (Maslow 1943, p. 5). In line with our understanding of safety perception, we reason that a presently experienced and further expected satisfaction of such physiological needs is a fundamental precondition for perceived safety.
2. *Safety needs.* Maslow's concept of safety refers to threats of personal security, health and financial security such as crime, murder, extreme weather conditions and severe illness but also unemployment as well as generally insufficient predictability and order (Maslow 1943). By classifying safety needs as the needs category of the second highest order, Maslow's theory highlights the high importance for humans to feel safe and the strong motivation to attain a state of perceived safety: "Practically everything looks less important than safety (even sometimes the physiological needs which being satisfied, are now underestimated). A man, in this state, if it is extreme enough and chronic enough, may be characterized as living almost for safety alone" (Maslow 1943, p. 6). We go beyond Maslow in arguing that safety perception requires the satisfaction of further basic human needs.
3. *Love needs.* The needs category of love, affection and belongingness includes the need for stable relationships with one's family and friends as well as romantic relationships. Maslow (1943) stresses the importance of affection and belongingness by referring to the relation between social isolation and mental disorder. As the main emphasis of this chapter lies on the materiality of social relationships, we discuss social needs in general as well as the various negative effects of lacking social relationships in a later part of this section. At this point, we want to capture that affection and belongingness are fundamental human needs and—in line with our broad concept of safety—are preconditions for safety perception.
4. *Esteem needs.* Maslow identified a stable, high evaluation of oneself as a basic human need (Maslow 1943). "All people in our society (with a few pathological exceptions) have a need or desire for a stable, firmly based, usually high evaluation of themselves, for self-respect, or self-esteem, and for the esteem of others" (Maslow 1954, p. 90, as cited by Huizinga 1970). The esteem needs thereby include both a need for self-respect as well as for respect from others (Huizinga 1970). Based on the assumption that the dissatisfaction of all basic human needs results in actual harm (Baumeister 2012), we argue that perceived safety requires also the satisfaction of such needs that exceed mere physical integrity and predictability.

5. *Need for self-actualization.* Maslow differentiates between deficiency needs and growth needs, whereby the four previously introduced needs categories are deficiency needs. As soon as the deficit (e.g., hunger, criminal threat, loneliness, lack of recognition) is gratified, the need is satisfied, and the person is no longer motivated to improve the situation. We would argue, instead: As soon as the deficit is gratified, the person feels safe. However, in the case of growth needs, the fulfillment of the need remains attractive regardless of how much it has already been satisfied (Huizinga 1970). Maslow introduced the need for self-actualization as a growth need: "the desire to become more and more what one is, to become everything that one is capable of becoming" (Maslow 1954, p. 92, as cited by Huizinga 1970).

Regarding the deficiency needs, we find it important to bear in mind that the point of fulfillment of these needs is in many cases not objectively defined. There are numerous factors that can influence individual aspiration levels of different needs. We believe that besides innate physiological needs such as hunger and thirst, aspiration levels can be socially determined such that others influence the point of fulfillment of one's needs as well as how that fulfillment can be achieved: What is an attractive income? What is a friend and how many of them do I need? What sort of romantic relationship makes me happy? What kinds of achievements are seen as valuable? What indicates personal development? (…) Cultures, societies, families, circles of friends, peers, schools, universities, companies, etc. suggest different answers to these questions and can thereby shape (or confuse) individual aspiration levels. Depending on the culture, region, family, and zeitgeist one was born in, the concept of valuable relationships, success, personal development, etc. might be completely different. We argue that the existence of certain needs, the perception of whether these needs are fulfilled or not, and convictions on how these needs can be fulfilled are to a large extent socially mediated beliefs.

In addition to Maslow's hierarchy of needs, psychological research has identified further basic needs and motives that are of fundamental importance to humans and therefore fundamentally influence human well-being and behavior. For example, self-determination theory (Deci and Ryan 1985) suggests the existence of three universal human needs that predict human well-being across cultures: (1) autonomy (i.e., being a causal agent regarding what activities a person engages in and how), (2) competence (i.e., perceived effectiveness in one's activities), and (3) social relatedness (i.e., perceived closeness to others; Deci and Ryan 2008a; Sheldon et al. 2001). Environments (e.g., at work, in hospitals, or at home) that facilitates the satisfaction of these three basic needs motivate people effectively and are related to positive psychological and behavioral outcomes, whereas environments that do not allow for the satisfaction of these needs reduce both motivation and general human well-being (Deci and Ryan 2008b). We claim that these three motives overlap at least to a certain extent with the motives introduced by Maslow (especially his love needs, esteem needs and the need for self-actualization). Further psychological needs have been identified, such as a need for pleasurable stimulation (as suggested by Epstein's cognitive-experiential self-theory; see Sheldon et al. 2001), the human desire and need for control in terms of explainability, predictability, and influenceability (Frey and Jonas 2002; Streicher et al. 2012) such as for appreciation, fairness and meaningfulness (Frey et al. 2011), and many others.

We believe that all these basic human needs are to a certain extent essential for the perception of structure, predictability, self-confidence and thereby safety. We claim that feeling safe depends not only on physical integrity but also on many further factors such as whether one has positive and stable social relationships with others, whether one feels fairly and respectfully treated, whether one is satisfied with oneself and able to develop in a desired way, whether one has influence on a given situation and is able to anticipate the ongoing development, whether one has a certain freedom in what to do and how to do it, etc. In the following we want to consider more specifically the basic human need that was described by Maslow as *love needs* and by Deci and Ryan as *social relatedness*: The need for social belongingness.

### *4.2.2 Need to Belong—Why We Need Relationships to Feel Safe*

Need to belong theory (Baumeister and Leary 1995) says that humans have a fundamental need to belong with others and thus establish and maintain social relationships. More specifically, humans seek stable, caring relationships as well as regular interactions. Baumeister (2012) stresses the high priority of the need to belong with the observation that humans often take physical risks in order to impress others. Riding a motorbike without a helmet, smoking cigarettes, and sunbathing can be seen as examples for such behaviors that illustrate that the need to belong can outperform other fundamental needs (Baumeister 2012). We want to discuss such obviously risky behaviors to explain the safety perspective we take in the present chapter: Riding a motorbike without a helmet, smoking cigarettes and sunbathing are all behaviors that are generally seen as unsafe. Of course, one can argue that people engage in these behaviors simply because of a certain need for pleasurable stimulation (see Sheldon et al. 2001). However, we believe that these behaviors can also be explained through the *social safety* they promise: While these behaviors are indeed unsafe as far as their effect on one's physical health, we argue that under certain conditions, such behaviors might be safe options in relation to one's need to belong: Smoking might be attractive in certain contexts as a safe strategy to impress peers or even to make or keep friends. Comparatively, riding a motorbike without a helmet and sunbathing might be effective strategies to make a certain impression on others and thereby to satisfy one's need to belong.

The great importance of belongingness for human well-being became apparent through many experimental studies demonstrating the drastic effects of social exclusion. For example, scholars induced social exclusion in groups by having participants in a group choose another group member as his or her partner, and then telling certain individuals that he or she had not been chosen as a partner by any other group member (Baumeister 2012). Other experiments used the context of a ball game where the group members suddenly stop throwing the ball to a particular (socially excluded) participant (see Williams and Jarvis 2006). People's immediate reactions to threats to their need to belong usually are negative affect (i.e., socially excluded persons perceive negative emotions) as well as lower self-esteem (i.e., social exclusion harms a person's subjective evaluation of his or her own self; Smart Richman and Leary 2009). However, further responses to exclusion vary over a broad range from antisocial behavior (Warburton et al. 2006) to socially avoidant behavior and even prosocial behavior (Lakin and Chartrand 2003).

Excluded persons simultaneously experience the following three motives, given here in order of strength of influence: First, excluded persons perceive a heightened desire for social relations—either with the rejecting person(s) or with others who can provide social relatedness. Second, excluded persons feel angry and urged toward aggressive, antisocial responses with the aim of self-protection and harm toward the source of rejection. Third, socially excluded persons can be motivated to isolate themselves from social contacts to avoid further rejection and the related pain (Smart Richman and Leary 2009). A set of factors determines which of the three motives dominates people's responses to social rejection (e.g., are there alternative relationships available? How pervasive is the exclusion? Is the exclusion perceived as unfair?). Thus, as social exclusion threatens a person's need to belong, there is a certain willingness toward cooperation and thus approaching the group. However, when people do not see a way to reintegrate into the group, they use antisocial responses and/or withdrawal as strategies of self-protection.

Interestingly, culture also seems to affect how people respond to social exclusion. People from collective cultures such as Turkey, India and China were shown to be less affected by social exclusion than people from individualistic cultures such as Germany and the United States. People from individualistic cultures tend to respond more strongly to exclusion than participants from collective cultures, with increased heart rates, higher levels of negative mood as well as more prominent antisocial and avoiding behavior (Pfundmair et al. 2015a; b).

The effects of social exclusion demonstrate the enormous importance of social belongingness for humans. Being accepted by the people who surround us seems to be a fundamental condition for feeling safe. Another research stream indicating the high significance of social relatedness for safety perception is that around attachment theory (Ainsworth and Bowlby 1991; Bretherton 1992). Empirical work in the field of attachment theory has indicated that the quality of a child's early attachment to his or her parents (or another primary caregiver) can have remarkable effects on the child's social and emotional development as well as present and future emotions and behaviors (Grossmann and Grossmann 2014). In that context, a key feature of parenting is described as "the provision by both parents of a secure base from which a child or an adolescent can make sorties into the outside world and to which he can return knowing for sure that he will be welcomed when he gets there, nourished physically and emotionally, comforted if distressed, reassured if frightened" (Bowlby 2005, p. 11). This first section of this chapter shall illustrate the importance of social relations for human well-being in general and safety perception in particular. Being accepted and liked by the persons who surround us is of such importance for us that it has a material impact on our everyday judgment formation and behavior. In the following, we discuss further how social factors influence our perceptions and judgments on how safe or risky a certain situation is.

## 4.3 Social Influences on Safety Judgments and Decisions

We have illustrated social belongingness as a basic human need. It is therefore not surprising that we find ourselves well-embedded in social relations in most parts of our daily lives: We usually live in close interaction with our families and friends, work in teams or cultivate business relationships, and also seek contact with others in our leisure time such as when we participate in sports, engage in cultural events or use social media. As our everyday lives are dominated by social relations and interactions, it is no surprise that we consider what others say and do when making safety-related judgments and decisions. When we face questions like *is it safe to go swimming today, do I need accident insurance for my children, should I wear a helmet when riding my bike to work, is this a safe trip,* or *should I buy shares of that company,* we naturally consider how others deal with these issues. Thus, we might have a look at the sea and check if there are other people swimming, ask our friends if they have accident insurance for their children (and why), consider whether our colleagues wear a helmet when they come to work by bike, search for reports of travel experiences on the internet, or discuss investment options with our colleagues.

Per social perception theory (Bruner and Postman 1948), perception is a process of constant testing of the assumptions people have about their surroundings. People consult the behavior of significant others to verify or disprove their assumptions, such as about how safe a certain situation, technology or activity is. It is also a basic concept of social comparison theory (Festinger 1954) that people evaluate their opinions and assumptions by comparing them to those of others. Humans want to make correct judgments and decisions and compare their assumptions with the statements and behaviors of others as a sort of social validation strategy. It is a main element of dissonance theory (Festinger 1962), that humans seek the support of like-minded people when uncertain about their judgments or decisions. Especially in uncertain, fearful situations, one seeks to affiliate with others (see Schachter's affiliation experiments; Schachter 1959). Thus, people base their safety-related assumptions and judgments (such as whether a certain food is poisonous) to a high degree on the statements and behaviors of others. In the following, we discuss psychological findings on how social factors in general and groups in particular affect our everyday safety judgments and decisions. We go on to discuss how our personal safety perception depends on the behaviors and expressions of the people who surround us.

### 4.3.1 Group Conformity—Groups Make Us Think More Equally

One of the earliest experiments demonstrating how dramatically human judgment can be affected by social factors was conducted by Sherif (1935). In his experiment, participants were asked to estimate the distance covered by a (seemingly) moving point of light in a completely darkened room. Of course, without any three-dimensional context information, there was no indication that would justify a particular assessment: It was absolutely arbitrary whether participants estimated the distance as rather short (because the light was assumed to be close to the participant) or rather distant (because the light was assumed to be far away from the participant, see Fig. 4.2). It is therefore no surprise that Sherif's participants expressed totally different distance estimations. Sherif's remarkable finding, however, was that as soon as he asked his participants to estimate the distance covered by the light in a group setting, the expressed estimations became more equal. Notably, the higher convergence of the estimated distances remained when participants were asked to re-estimate the distance individually afterwards. The experiment shows that groups build certain common norms and that these norms affect both our group and individual judgments in such a way that the convergence of individual judgments increases. Sherif's participants implicitly agreed on a shared understanding of how far the light was away. That norm affected not only the distance estimates within the group but also the subsequent individual estimations (Erb et al. 2002; Sherif 1935).

Many of the safety-related questions that we face in our everyday lives can be of similar difficulty to the ambiguous task in Sherif's experiment. Common norms of what is generally perceived as safe or risky can facilitate answering such difficult questions. Common assumptions within groups can affect the safety judgments of individuals regardless of whether the group has a right or wrong understanding of the situation. It has been shown that the risk categories of technological, chemical and environmental risks tend to be in many cases overestimated (Renn 2014). When a group collectively assumes that new technologies generally entail high risk, this common norm can also influence the safety perceptions of individual group members (as indicated by Sherif's experiment).

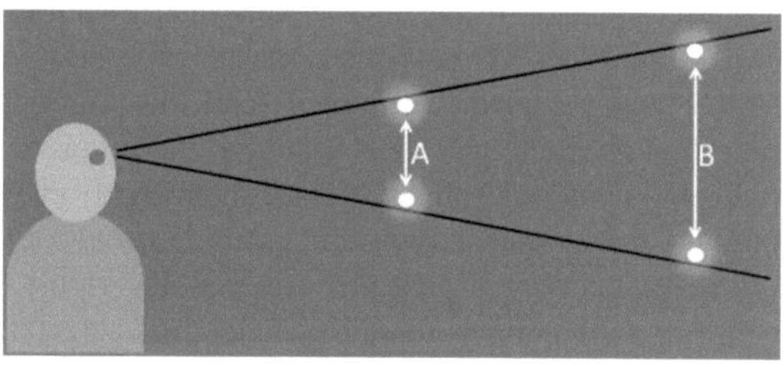


**Fig. 4.2** Illustration of Sherif's (1935) experiment demonstrating the effect of group norms

**Fig. 4.3** Illustration of Asch's (1951) test material for demonstrating the effect of group conformity

While the experiment conducted by Sherif (1935) demonstrated the effects of group norms on both group and individual judgments when faced with a highly ambiguous task, it was later demonstrated that group norms also affect human judgments in apparently non-ambiguous situations. In 1951, Asch conducted an experiment in which participants were asked to indicate which of three lines were of the same length as a reference line (see Fig. 4.3 for an illustration of the test measures). Participants were asked to make multiple judgments of that type in groups of six persons. For every task, each of the group members was asked to express a solution one after another. The twist in Ash's experiment was that all but one participant in each group were actors instructed to express a solution that was wrong but consistent amongst all actors. As a consequence, many participants adapted their answers to the obviously wrong judgments of their peers. Asch's experiment provides drastic evidence for the phenomenon of group conformity: Even though the correct answer was clearly recognizable, many participants expressed obviously wrong solutions that conformed to the solutions suggested by the other group members. The inclination to act in conformity to the group was strong enough that participants expressed solutions which they knew were wrong (Erb et al. 2002).

A real-world phenomenon that reflects Asch's findings and can be currently observed in (middle European) winter skiing resorts is that up to 95% of skiers and snowboarders wear helmets, whereas only a few years ago, almost nobody wore skiing helmets (Focus online 2015). While the initial cause of this impressive trend reversal is presumably mostly attributed to the success of skiing equipment producers in making helmets appear more attractive, we argue that a changed common norm within the group of skiers and snowboarders also played a role in this self-reinforcing process: When a majority of skiers and snowboarders adopted wearing helmets, it became *normal* and thus *right* to do so. Not wearing a helmet accordingly came to be seen as abnormal, reckless behavior that would provoke critical questions. This is the opposite of a few years ago, when not wearing a helmet was perceived as normal (and thus right) and wearing one would have provoked skeptical glances. Of course, the actual risk of skiing and snowboarding (with and without a helmet) did not significantly change during this timeframe. The only thing that changed was the group norm about wearing helmets.

Asch's experiment and the winter skiing resort example demonstrate how the behavior of majorities can affect individual perceptions of how safe a situation is

(*If everybody wears a helmet not wearing one must be risky!*). And it is not only majorities such as in Asch's experiments but also minorities that can have an impact on our judgment formation. The experiments of Serge Mosovici on the influence of minorities demonstrate that the latter can have influence on other group members' judgments and decisions, especially if their behavior is consistent within the minority. Minorities can thereby provoke cognitive conflicts and induce others to rethink their current positions and judgments (Moscovici and Faucheux 1972; Moscovici et al. 1969). When considering how the high willingness to wear a helmet amongst skiers and snowboarders might be created in other contexts such as amongst cyclists or roller-skaters, the consistent wearing of a helmet by smaller groups can have the effect of convincing others to wear one also, simply by provoking critically questioning of the current group norms.

So far we have pointed out that groups can significantly affect the perceptions, beliefs and behaviors of their members. How can this remarkable phenomenon be explained? People identify themselves not only as individuals but also as group members. This is the core of the so-called social identity theory introduced by Tajfel and Turner (1986). Social identity describes an individual's self-conception based on perceived membership of social groups as well as on how this membership is evaluated (Ellemers and Haslam 2012). Individuals quickly learn what kind of behaviors are expected and desired within a group of which they consider themselves to be a member. What happens next is often described as the phenomenon of *depersonalization* or *self-stereotyping:* The person starts redefining her current individual perceptions and beliefs in a way that closely aligns with what is assumed to be the perceptions and beliefs of the group. The person *self-stereotypes*, moving from her individual perspective to the (assumed) perspective of the group. Characteristics of the group become characteristics of the individual. Further, the perception of other ingroup members as similar to oneself enhances the willingness to trust their views and follow their examples. Being part of the same group makes people perceive themselves as relatively similar. Because of the perceived similarity, group members tend to think that they *ought* to have similar views and *ought* to agree easily and quickly. Thus, group members express what they believe is expected from the group and confirm what others in the group suggest (Turner and Reynolds 2011).

### 4.3.2 Groups Make Extreme Judgments

We have seen that the mere existence of a group has a certain effect on our perceptions, judgments and decisions. Because safety-related questions are often complex, we often seek the advice of others or aim to discuss certain issues within groups. In fact, when making important safety decisions such as whether to buy insurance, travel to a certain country or take a certain vaccination, most people seek advice from family, friends or professionals. Professional safety decisions, such as on public or corporate safety, are also rarely taken alone by individual persons.

Rather, there are special councils, committees and working groups that discuss the issue in detail and then reach an agreement. In the following, we illustrate that such group discussions and decisions entail the risk of biased judgments.

In 1961, Stoner asked his participants to judge the risk of various rather safe and rather risky options. He first asked each of the participants for an individual risk estimation and then had his participants discuss and judge the options in a group. Finally, he asked his participants again individually for their judgments. Stoner thereby discovered an astonishing phenomenon: Both the group estimations and the average of all individual estimations after the discussion were riskier than the individual judgments before the discussion. Something had happened during the group discussion that convinced his participants of a higher risk judgment. The discovered phenomenon was later called *risky shift* and describes how group discussions can have the effect of a shift towards more risky decisions (Stoner 1968).

The risky shift can be considered as a special case of the more general phenomenon of group polarization, which has been described by Moscovici and Zavalloni (1969): During group discussions, both group judgments but also average individual judgments shift in the direction of the initial average tendency of the group members before the group discussion. If the group members' initial average tendency was riskier, a group discussion amplifies that risky tendency with the result of a riskier group judgment (as it was the case in Stoner's original experiment). However, the effect can also cause a shift in the other direction: If the group members' initial average tendency was more cautious before a group discussion, the group discussion enhances that initial tendency, with the result of an even more cautious group judgment and average individual judgment. This form of group polarization is called *caution shift* (Schulz-Hardt et al. 2002; see Fig. 4.4 for an illustration of risky shift and caution shift).

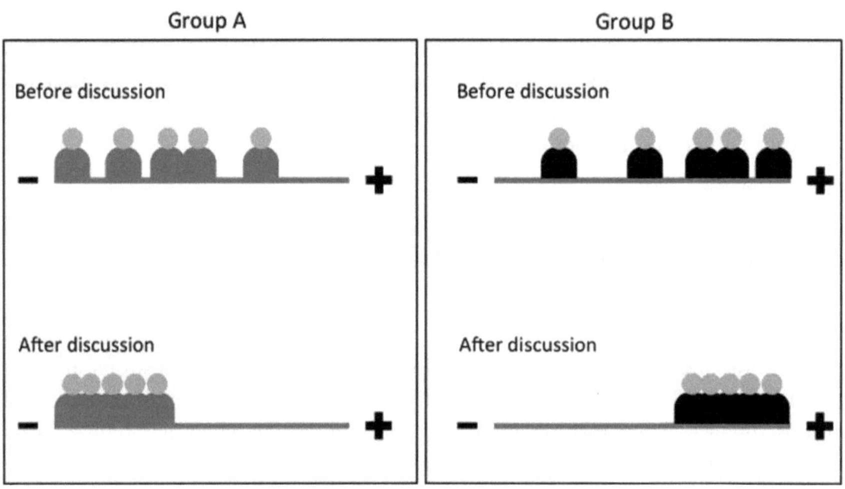


**Fig. 4.4** Illustration of caution shift (A) and risky shift (B)

How is this shift in people's risk judgments possible? We have learned that people are particularly sensitive to expected behaviors within groups. Because humans seek acceptance within groups, they are willing to adapt their judgments and behaviors to these expectations (Turner and Reynolds 2011). In group discussions, the tendency to express statements that conform with the group norms can lead to group polarization: Many group members express arguments for the same position or opinion and thereby convince the group of a more and more extreme version of the initial position. Furthermore, in order to be considered an active group member, it is an effective (and popular) strategy of individual group members to express statements that are similar to these of the other group members but somewhat more extreme. This, of course, has an amplifying effect on the group's judgment such that the assumptions and judgments within the group become more and more extreme (for an overview, see Sunstein 2005).

Groups are heavily influenced in their judgments by their initial tendencies. When authorities or thought leaders express their opinions or assumptions in the beginning of a group discussion, it is likely that the group will adopt that view as a starting point, and many group members will express arguments speaking for the same position. In the end, the group is likely to come to a conclusion that is similar but somehow more extreme than what was initially suggested by the thought leader. Under the heading of the *yale attitude change approach* studies, Carl Hovland conducted several experiments with the aim of identifying (amongst others) personal characteristics that allow persons to influence other people's attitudes. The studies indicate that both the credibility (which comprises expertise and trustworthiness) and the attractiveness of a communicator can significantly influence people's attitudes (Hovland and Weiss 1951). Thus, groups tend to adopt the opinion of persuasive individuals and even intensify that view through group discussions. Because being accepted and loved by the people who surround us is of such fundamental importance, people are willing to adapt to groups' norms and expectations. Group conformity and group polarization are two examples for how such social effects can influence a person's safety judgments and decisions.

## 4.4 Outlook

The present chapter is an attempt to illustrate and summarize how social factors influence human safety perception. We have suggested that safety perception depends on social factors in multiple ways. In particular, we have illustrated that social relatedness is an important condition for feeling safe. Also, we have seen that groups influence one's judgments and decisions on safety-related questions. Figure 4.5 provides an illustration of some of the most important content and relationships we suggest in this chapter.

We have introduced perceived safety as a state that exists beyond any fundamental experienced or feared threats to a person. Thus, we argue that feeling safe relies to a large extent on the satisfaction of the most basic human needs. While we

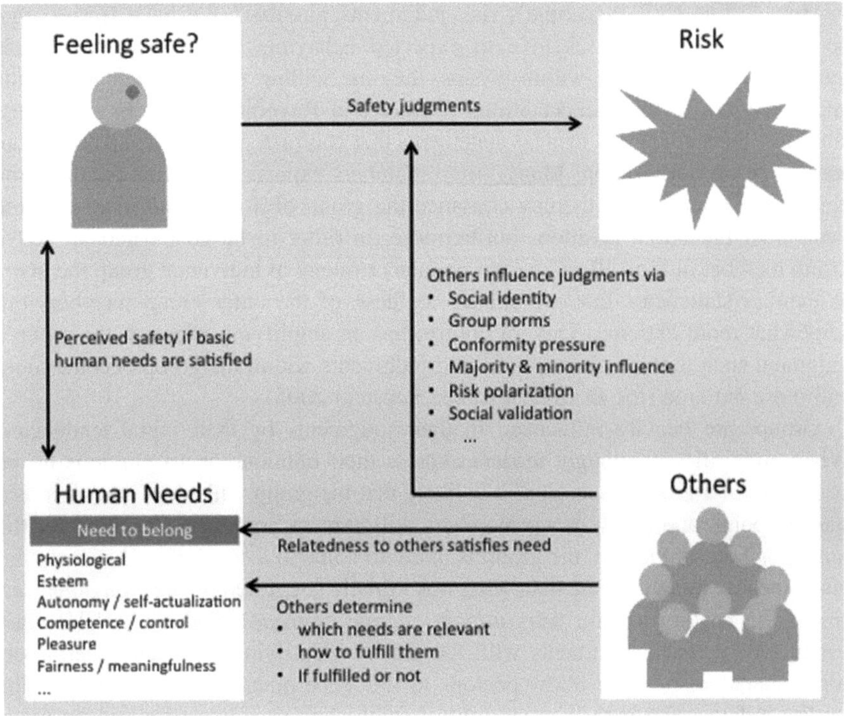


**Fig. 4.5** Illustration of social factors of safety perception

argue that all basic human needs relate to the perception of safety, we lay particular emphasis on the need for social belongingness. With reference to need to belong theory, we argued that someone will only entirely feel safe when the need for social belongingness is satisfied. We have illustrated the intensity of that need with the drastic, mostly negative effects of social exclusion that have been demonstrated repeatedly by scholars. Given the strong pain people perceive when not being accepted by the persons and groups who surround them, it is even possible to explain seemingly risky behaviors (we have mentioned smoking cigarettes, riding a motor bike without a helmet, and sunbathing as examples) via the strong motivation to be accepted by others.

Making people feel safe is (or at least should be) a basic goal of decision-makers in all parts of private, professional and public life. How we can ensure that people feel safe in their families, at their school, in their work places, in our cities, etc. should be a major concern. We suggest that considering basic human needs is a valuable and constructive perspective for effective improvements in all these contexts. In this chapter, we have laid particular emphasis on the need for social relatedness, and we do believe that facilitating social relatedness is an important means for the creation of perceived safety. However, a considerable number of further basic human needs need to be considered, such as the need for autonomy, competence, pleasure stimulation, control, fairness, and meaningfulness. To provide guidance for practical implementation, Frey's ethical leadership model might be a valuable starting point for practitioners who aim at increasing perceived safety. The model is a practice-oriented approach that is based on the principle of aligning leadership according to human needs such as meaningfulness, transparency, autonomy and participation, constructive feedback, stimulation, personal growth, and fairness (see Frey and Schmalzried 2012; Frey et al. 2012). We believe the model can provide valuable guidance for the implementation of practical interventions to make people feel safe.

The second part of the present chapter builds on the observation that important safety-related decisions are either made by groups (e.g., councils, committees, or working groups) or by individuals who implicitly consider the behaviors of others or explicitly ask for advice for the decisions they make. Thus, safety judgments and decisions depend on social factors. We have presented and discussed a number of classical psychological experiments demonstrating how individuals adapt their judgments, decisions, and behaviors to group norms. As soon as others are present, being accepted by those people can be more important than making an accurate judgment. Thus, safety decisions such as whether to wear a helmet, whether to buy a safe vs. an impressive car, or whether to cancel a sports event due to questionable weather can be biased by our beliefs on what others expect. We have discussed how common norms, such as within a certain culture, company or family, affect what is perceived as safe and what is perceived as not safe.

Within a discussion group, the tendency to state facts, opinions and arguments that conform with the group's overall position can lead to the phenomenon of group polarization: Groups tend to make extreme judgments and decisions and thus agree on an understanding that a certain situation or opportunity is either very risky or very safe. Therefore, it should generally be questioned when closed groups quickly come up with clear and easy responses to complex questions. An approach that can help reduce such social effects in discussion groups is to structure the process of group discussions and decisions. For instance, it can increase the quality of a decision to ask different persons about their perspective individually and anonymously and then decide based on the collected opinions and arguments in a second step (see Lermer et al. 2014).

Our chapter illustrates that humans are social beings in need of belongingness and in search of approval from others for their judgments and decisions. Some of the presented findings may create the impression that groups generally bias safety estimations and that one should therefore better make important safety decisions alone. It is important for us to emphasize that the social validation we seek for our daily judgments and decisions is in the majority of cases a very functional strategy. Adapting one's behavior to what works for others is efficient and leads in most cases to good results. However, we also presented examples for how such group effects can worsen the quality of a safety judgment (e.g., when a group uncritically adopts the view of a thought leader and reinforces that view through numerous consistent statements in a group discussion). Being aware of these mostly unconscious social effects on safety perception is a precondition for consciously making safety-adequate decisions, both in groups and individually.

To finish, we want to stress that the best safety decision is not always the most conservative one. An overly overt focus on safety bears the risk of constraining freedom and can impede experiments, new ideas, and innovation. We have illustrated that our conception of safety in general and also the many related constructs (e.g., human needs, the respective aspiration levels, and whether and how they are fulfilled) are to a large extent determined by what other people say and do. Bearing in mind that existing safety convictions are sometimes not necessarily based on evidence or reason, this chapter encourages critical questioning of the status quo of safety beliefs and behaviors.

## References

Ainsworth, M. S., & Bowlby, J. (1991). An ethological approach to personality development. *American Psychologist, 46,* 333–341.

Asch, S. E. (1951). Effects of group pressure upon the modification and distortion of judgment. In H. Guetzkow (Ed.), *Groups, leadership and men* (pp. 177–190). Pittsburgh: Carnegie Press.

Baumeister, R. F. (2012). Need-to-belong theory. In P. A. M. Van Lange, A. W. Kruglanski, & E. T. Higgins (Eds.), *Handbook of Theories of Social Psychology* (Vol. 2, pp. 121–140). Thousand Oaks: Sage.

Baumeister, R. F., & Leary, M. R. (1995). The need to belong: desire for interpersonal attachments as a fundamental human motivation. *Psychological Bulletin, 117,* 497–529.

Bowlby, J. (2005). *A secure base: clinical applications of attachment theory* (Vol. 393). London: Taylor & Francis.

Bretherton, I. (1992). The origins of attachment theory: John Bowlby and Mary Ainsworth. *Developmental Psychology, 28,* 759–775.

Bruner, J. S., & Postman, L. (1948). An approach to social perception. In W. Dennis (Ed.), *Current trends in social psychology* (pp. 71–118). Oxford: U. Pittsburgh Press.

Deci, E. L., & Ryan, R. M. (1985). *Intrinsic motivation and self-determination in human behavior.* New York: Plenum Press.

Deci, E. L., & Ryan, R. M. (2008a). Self-determination theory: a macrotheory of human motivation, development, and health. *Canadian Psychology, 49,* 182–185.

Deci, E. L., & Ryan, R. M. (2008b). Facilitating optimal motivation and psychological well-being across life's domains. *Canadian Psychology, 49,* 14–23.

Ellemers, N., & Haslam, S. A. (2012). Social identity theory. In P. A. M. Van Lange, A. W. Kruglanski, & E. T. Higgins (Eds.), *Handbook of theories of social psychology* (pp. 379–398). Thousand Oaks: Sage.

Erb, H. P., Bohner, G., Rank, S., & Einwiller, S. (2002). Processing minority and majority communications: the role of conflict with prior attitudes. *Personality and Social Psychology Bulletin, 28,* 1172–1182.

Festinger, L. (1954). A theory of social comparison processes. *Human Relations, 7,* 117–140.

Festinger, L. (1962). *A theory of cognitive dissonance* (Vol. 2). Stanford: Stanford University Press.

Frey, D., & Jonas, E. (2002). Die Theorie der kognizierten Kontrolle. In D. Frey & M. Irle (Eds.), *Theorien der Sozialpsychologie Band 3: Motivations-, Selbst- und Informationsverarbeitungstheorien.* Bern: Verlag Hans Huber.

Frey, D., Oßwald, S., Peus, C., & Fischer, P. (2011). Positives Management, ethikorientierte Führung und Center of Excellence—Wie Unternehmenserfolg und Entfaltung der Mitarbeiter durch neue Unternehmens-und Führungskulturen gefördert werden können. In M. Rinlstetter, S. Kaiser, & G. Müller-Seitz (Eds.), *Positives management* (pp. 239–270). Wiesbaden: Gabler Verlag.

Frey, D., & Schmalzried, L. K. (2012). Das Modell der ethikorientierten Führung. In F. Zschaler, S. Meck, & J. Kleine (Eds.), *Finethikon Jahrbuch für Finanz- und Organisationsethik* (Vol. 2, pp. 61–92). Stuttgart: Steinbeis-Edition.

Frey, D., Streicher, B., & Aydin, N. (2012). Center of Excellence Kulturen sowie professionelle ethikorientierte Führung als Voraussetzung für ökonomischen Erfolg. In S. Grote (Ed.), *Die Zukunft der Führung* (pp. 235–253). Berlin: Springer.

Focus Online (2015). http://www.focus.de/familie/freizeit/stylish-muss-er-sein-skihelmpflicht_id_1880452.html.

Grossmann, K., & Grossmann, K. E. (2014). *Bindungen—das Gefüge psychischer Sicherheit.* Stuttgart: Klett-Cotta.

Hovland, C. I., & Weiss, W. (1951). The influence of source credibility on communication effectiveness. *Public Opinion Quarterly, 15,* 635–650.

Huizinga, G. H. (1970). Maslow's need hierarchy in the work situation. Doctoral dissertation, TU Delft, Delft University of Technology.

Kahneman, D., & Tversky, A. (1979). Prospect theory: an analysis of decision under risk. *Econometrica: Journal of the Econometric Society, 47,* 263–291.

Lakin, J. L., & Chartrand, T. L. (2003). Using nonconscious behavioral mimicry to create affiliation and rapport. *Psychological Science, 14,* 334–339.

Lermer, E., Streicher, B., Eller, E., & Sachs, R. (2014). Psychological influences II: risk assessment in groups. *Munich Re Discussion Paper,* 1–4.

Lewin, K., & De Rivera, J. (1976). *Field theory as human-science: contributions of Lewin's Berlin group.* New York: Gardner Press.

Maslow, A. H. (1943). A theory of human motivation. *Psychological Review, 50,* 370–396.

Moscovici, S., & Faucheux, C. (1972). Social influence, conformity bias, and the study of active minorities. *Advances in Experimental Social Psychology, 6,* 149–202.

Moscovici, S., Lage, E., & Naffrechoux, M. (1969). Influence of a consistent minority on the responses of a majority in a color perception task. *Sociometry, 32,* 365–380.

Moscovici, S., & Zavalloni, M. (1969). The group as a polarizer of attitudes. *Journal of Personality and Social Psychology, 12*(2), 125–135.

Oxford Dictionaries (2015). http://www.oxforddictionaries.com/definition/english/safety.

Pfundmair, M., Aydin, N., Du, H., Yeung, S., Frey, D., & Graupmann, V. (2015a). Exclude me if you can—cultural effects on the outcomes of social exclusion. *Journal of Cross-Cultural Psychology, 46,* 579–596.

Pfundmair, M., Graupmann, V., Frey, D., & Aydin, N. (2015b). The different behavioral intentions of collectivists and individualists in response to social exclusion. *Personality and Social Psychology Bulletin, 14,* 363–378.

Renn, O. (2014). *Das Risikoparadox—Warum wir uns vor dem falschen fürchten.* Frankfurt am Main: Fischer Verlag.

Schachter, S. (1959). *The psychology of affiliation: experimental studies of the sources of gregariousness (No. 1).* Palo Alto: Stanford University Press.

Schulz-Hardt, S., Greitmeyer, T., Brodbeck, F., & Frey, D. (2002). Sozialpsychologische Theorien zu Urteilen, Entscheidungen, Leistung und Lernen in Gruppen. In D. Frey & M. Irle (Eds.), *Theorien der Sozialpsychologie.* Bern: Hans Huber Verlag.

Sheldon, K. M., Elliot, A. J., Kim, Y., & Kasser, T. (2001). What is satisfying about satisfying events? Testing 10 candidate psychological needs. *Journal of Personality and Social Psychology, 80,* 325–339.

Sherif, M. (1935). A study of some social factors in perception. *Archives of Psychology (Columbia University), 187,* 60.

Smart Richman, L., & Leary, M. R. (2009). Reactions to discrimination, stigmatization, ostracism, and other forms of interpersonal rejection: a multimotive model. *Psychological Review, 116,* 365–383.

Stoner, J. (1961). *A comparison of individual and group decisions involving risk* (Master's thesis). https://dspace.mit.edu.

Stoner, J. (1968). Risky and cautious shifts in group decisions: the influence of widely held values. *Journal of Experimental Social Psychology, 4,* 442–459.

Streicher, P. D. B., Frey, D., & Graupmann, V. (2012). Fairness bei Veränderungsprozessen. In L. von Rosenstiel, E. von Hornstein, & S. Augustin (Eds.), *Change management praxisfälle* (pp. 361–375). Berlin: Springer.

Sunstein, C. R. (2005). *Why societies need dissent.* London: Harvard University Press.

Tajfel, H., & Turner, J. C. (1986). The social identity theory of intergroup behavior. In S. Worchel & G. W. Austin (Eds.), *The psychology of intergroup relations* (pp. 7–24). Chicago: Nelson Hall.

Turner, J. C., & Reynolds, K. J. (2011). Self-categorization theory. *Handbook of Theories in Social Psychology, 2,* 399–417.

Tversky, A., & Kahneman, D. (1973). Availability: a heuristic for judging frequency and probability. *Cognitive Psychology, 4,* 207–232.

Wahba, M. W., & Bridwell, L. G. (1976). Maslow reconsidered: a review of research on the need hierarchy theory. *Organizational Behavior and Human Performance, 15,* 212–240.

Warburton, W. A., Williams, K. D., & Cairns, D. R. (2006). When ostracism leads to aggression: the moderating effects of control deprivation. *Journal of Experimental Social Psychology, 42,* 213–220.

Williams, K. D., & Jarvis, B. (2006). Cyberball: a program for use in research on interpersonal ostracism and acceptance. *Behavior Research Methods, 38,* 174–180.

# Chapter 5
# Psychological Perspectives on Perceived Safety: Zero-Risk Bias, Feelings and Learned Carelessness

**Martina Raue and Elisabeth Schneider**

**Abstract** In this chapter, we introduce three common decision-making strategies that humans apply in situations of risk and uncertainty. Due to cognitive limitations, human beings often simplify complex decisions and use heuristics. People strive for safety, but tend to overweigh the value of zero risk in comparison to very small risks. Choosing the zero-risk solution is a heuristic that reduces complexity by eliminating the need to weigh statistical information, but may result in suboptimal decisions, which has been termed the *zero-risk bias*. Another strategy is rooted in the way humans process information. According to dual-process theories, information is processed intuitively (System 1) or analytically (System 2). Intuitive reactions, including affect and emotions, usually precede and often override analytical (cognitive) evaluations. The *affect heuristic* states that people judge risk information based on subtle feelings of positivity or negativity. A good feeling can therefore result in perceived safety despite diverging statistical information. Finally, one's attitude toward risks may be acquired through certain learning experiences. People who engage in risky behavior without encountering negative consequences may conclude that 'everything is fine and will remain fine,' which has been termed *learned carelessness*. Advantages and disadvantages of these strategies as well as practical implications, including decision aids and nudges, are discussed.

**Keywords** Certainty effect · Prospect theory · Affect heuristic · Selective information search · Unrealistic optimis · Illusion of control

---

Martina Raue and Elisabeth Schneider contributed equally to this chapter.

---

M. Raue (✉)
AgeLab, Massachusetts Institute of Technology, 77 Massachusetts Avenue, Cambridge, MA 02139, USA
e-mail: maraue@mit.edu

E. Schneider
Department of Psychology, Ludwig Maximilian University of Munich, Leopoldstr. 13, 80802 Munich, Germany



M. Raue et al. (eds.), *Perceived Safety*, Risk Engineering,
https://doi.org/10.1007/978-3-030-11456-5_5

## 5.1 Introduction

Most people do not like uncertainty and prefer a reliable and predictable world; in the latter case they feel safe (e.g., Van den Bos 2009; van den Bos et al. 2005). However, in real life we rarely encounter situations that offer 100% certainty or 100% safety—there is almost always a small chance, even in the safest situations, that something will go wrong. For example, most people feel quite safe when driving, although traffic accidents are a quite frequent cause of death (Statistisches Bundesamt 2015). This perceived safety while driving may stem from the fact that most drivers perceive themselves as more skillful and less risky than the average driver (Svenson 1981). In general, people tend to underestimate personal risks and think that others are more likely to be affected by potential harm. Similarly, most people are optimistic when it comes to their own longevity or success in comparison to others. This so-called *optimistic bias* is especially prominent when the risk is low, when people feel in control (e.g., when driving) and have not had negative experiences with a potential hazard (Weinstein 1989). Nevertheless, bad things might happen and, thus, some uncertainty remains and safety becomes a matter of perception. Perceived safety is especially relevant in situations that require judgments or decisions: *Should I drive or take the train? Is it safe to cross the street? Which medical treatment should I choose?*

Perceived risk varies between individuals due to their personalities, situational circumstances and experiences (e.g., Chauvin 2018; Lauriola and Weller 2018; Slovic and Peters 2006; Visschers and Siegrist 2018). Judgments are made in social, environmental and situational contexts which may evoke certain thoughts, feelings, or behaviors. While the previous chapter (by Eller & Frey) has highlighted social influences on perceived safety (e.g., influences of majorities, social comparison), we will focus on situational influences and particularly consider the roles of limited cognitive capacities, feelings and experiences.

When making decisions, human beings are not able to consciously consider all available relevant information, which has been termed *bounded rationality* (Simon 1955). These limitations become especially apparent under time pressure and with increasing complexity of a task. For instance, when having to make a choice about a medical treatment, humans struggle to weigh pros and cons of a treatment based on given probabilities of success, failure, or side effects. To simplify decision-making in these situations, but still maintain a sufficient level of accuracy, people use heuristics. Many heuristics work through attribute substitution, meaning that a complicated question is substituted with a simpler one (see Raue and Scholl 2018 for an overview). For example, instead of asking what the safest treatment is, one could ask which treatment has zero side effects or a 100% success rate. In many cases, heuristics lead to sufficient judgments and may even outperform judgments that attempt to include all the relevant information available, but in other instances these simplifications result in distorted judgments, known as *biases* (Hart 2005; Simon 1955; Tversky and Kahneman 1974). Focusing on certainty, which is also known as the *zero-risk bias*, may not be in one's best interest in a medical decision, as the tradeoff may be a less successful treatment (in the case of targeting zero side effects) or severe side effects (in the case of targeting a 100% success rate). The zero-risk bias will be discussed in the first part of this chapter.

A patient could also simply choose the treatment that he or she *feels* most comfortable with. Relying on subtle feelings of goodness or badness is a common decision-making strategy called the *affect heuristic*, which will be discussed in the second part of this chapter on the feeling of safety. Finally, if one has taken a medication in the past and not experienced any side effects, it may also be possible that this person will worry less about side effects and potentially underestimate potential risks related to them. This will be discussed in the third part of this chapter on *learned carelessness*.

## 5.2 Zero-Risk Bias—A Bird in the Hand Is Worth Two in the Bush

"A bird in the hand is worth two in the bush"—This common saying reflects a bias that has been well-observed in psychological research: People tend to overestimate the value of certainty in comparison to very small or very high probabilities and even accept considerable drawbacks in order to achieve perceived safety (Allais 1953; Kahneman and Tversky 1979; Viscusi 1997). Economically, rational behavior is seen as "behavior that maximizes the value of consequences" (Hastie and Dawes 2001, p. 249). However, when we recall incidents in the financial markets or unfinished building projects like the Berlin airport in Germany, human behavior seems to be far away from rationality. We have to depart from the assumption that humans are rational agents who make decisions by evaluating the possible options against their order of preferences, which, in turn, are defined by an individual utility function.

Allais (1953) was one of the first who took this theoretically into account and demonstrated that people distort objective probabilities ("la deformation subjective des probabilities objectives", p. 508). He found that people tend to over- or underweigh extreme probabilities equal or close to zero or 100, respectively, and demonstrated a so-called *certainty effect*:

Do you prefer option A or option B?
Option A: Gain 100 million for sure.
Option B: Gain 500 million with probability 0.1
Gain 100 million with probability 0.89
Gain nothing with probability 0.01

People preferred to win 100 million for sure (Option A) over the uncertain lottery, where there was a very small chance (1%) of winning nothing (Option B). However, objectively, this contradicts utility maximization. According to an expected value principle which defines the expected value as the product of outcome and probability (EV = v * p), the expected value of option B (EV (B) = 500 * 0.1 + 100 * 0.89 + 0 * 0.01 = 139) is higher than that of option A (EV(A) = 100) (e.g., Ajzen 1991; Atkinson 1957; Fishbein and Ajzen 1975). The example illustrates a bias referred to as certainty effect or zero-risk bias: rejecting a rational option in favor of a safe but more unfavorable option (e.g., Baron et al. 1993; Kahneman and Tversky 1979; Ritov et al. 1993).

Similarly, the increase of a risk from 0% to 10% makes a much larger difference for people than an increase from 20% to 30%. This is called *diminishing sensitivity* and can also be found in consumer decisions: Often, one would not spend 500 € without further consideration. However, when purchasing a car for 15.000 €, 500 € for an extra gadget is often paid without much further thought. Figure 5.1 illustrates the phenomenon of diminishing sensitivity, the effect where people become more insensitive to changes in numbers the higher they get, which is the second principle of *Prospect Theory* (Kahneman and Tversky 1979; Tversky and Kahneman 1992). Prospect Theory describes how humans actually make decisions and challenges the assumption of a rational decision-maker. The first principle of prospect theory is that decisions always refer to a reference point, which can be the status quo (current situation) or an expected outcome (e.g., a pay raise). An improvement referred to the reference point is perceived as a gain, while a decline is perceived as a loss. The third principle is loss aversion, meaning that losses loom larger than gains. For instance, the pain (i.e., psychological cost) of losing 100 € is much higher than the joy (i.e., psychological utility) of winning 100 € (note the difference between losses and gains in subjective value in Fig. 5.1).

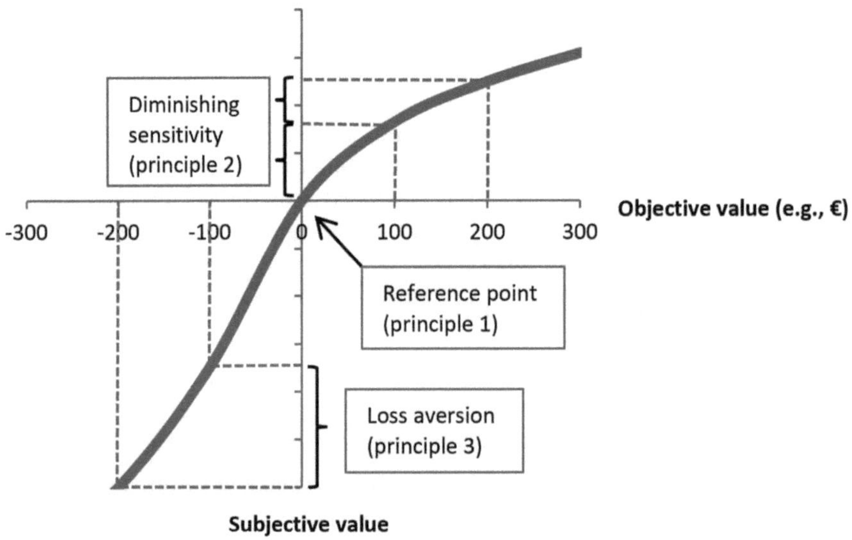


**Fig. 5.1** Value function of prospect theory (cf. Kahneman and Tversky 1979)

Both terms, certainty effect and zero-risk bias, describe in part the same phenomenon and are often used synonymously. The term zero-risk *bias* underlies the fact that the zero-risk option is detrimental compared to a risky option. Certainty effect seems to be the broader definition, in that the certain option is not implied to be the unfavorable decision. Since we focus on distortions in risky decision making which violate utility maximization, we use the term zero-risk bias.

In monetary terms, this bias is expressed by a *certainty premium* (Viscusi 1997). Hence, when we stated that people overweigh the value of certainty, this is to be taken literally, as demonstrated by the following study: Viscusi, Magat and Huber (1987) showed participants cleaning agents (insecticide and toilet bowl cleaner), each causing two possible toxic side effects (inhalation poisoning and skin/child poisoning). Both risks of side effects had an initial baseline of 15 cases per 10,000 bottles. Participants were informed about the products' prices and were then asked how much they would pay for the same-but-safer product (e.g., "15/10,000 inhalation poisonings and 0/10,000 skin/child poisoning", "5/10,000 inhalation poisonings and 5/10,000 skin/child poisonings", or "no side effects at all"). Depending on the type of poisoning and the product's price in general, reducing a side effect from 5/10,000 to 0/10,000 was perceived as up to three times more valuable than the reduction of the same side effect from 15/10,000 to 10/10,000 (note that the amount of risk reduction is similar in both cases). Moreover, there was a tendency to neglect the overall risk reduction: participants paid more for the complete elimination of one of the two toxic side effects (the risk of the other side-effect remained the same) than for the reduction of both risks to a certain rest-risk level (which resulted in a lower overall risk compared to the elimination of just one risk). To further illustrate this, consider the following example:

Assume that your firm is at risk of two losses that can both cause insolvency. Due to limited resources you can only intervene in one of these risks. The probability of the first risk occurring is 15%. Intervention A would reduce this risk to 5%. The probability of the second risk occurring is 5%. Intervention B would reduce this risk to 0%. Which intervention would you choose? (Eller et al. 2012).

The rational agent should choose intervention A, resulting in a lower overall risk (10.25% compared to 15% for intervention B). Note that both risks could occur so that the residual overall risk is $(A \times \bar{B}) + (\bar{A} \times B) + (A \times B)$. However, regardless of how much the risk was reduced, there were always participants who preferred the zero-risk option (intervention B) to the rest-risk option (intervention A) (Schneider et al. 2017). The same was found by Baron et al. (1993), who asked people to choose from three clean-up interventions for a waste deposit in order to reduce cases of cancer in their neighborhood. Of all participants, 11% rated the zero-risk option, which was the most objectively unfavorable of the three, as the best; and 42% did not rank it as the worst alternative. Using decision scenarios similar to the aforementioned example but from diverse decision domains (Blais and Weber 2006), research has shown that overall nearly 40% favor a zero-risk option (Schneider et al. 2017). Interestingly, people were highly sensitive to the decision domain (e.g., health, social, finance) but completely insensitive to changes in the overall risk, i.e., the favorability of the rest-risk option (like reducing the risk in intervention A from 60 to 5%).

The causes of this bias and whether it is driven by cognitive (e.g., limited cognitive capacity) or emotional (e.g., affect regulation) mechanisms is unclear. However, some key findings emerged from our own studies: First of all, the bias was unaffected by social demographics: men and women, the young and the old, students and businessmen, engineers and educators all showed the bias. This is contradictory to findings of Mather and colleagues (2012) who found that older adults showed a greater preference to receive sure gains and avoid sure losses in gambling tasks than younger adults. Second (and in line with the findings of Viscusi et al. 1987), the amount of overall risk reduction did not affect people's preference for the certain solution. A decline in preferences for the zero-risk option was only found when the overall risk reduction became very large (e.g., above 50%), however, smaller overall risk reductions did not affect participants' choice (Schneider et al. 2017). For example, if people could reduce a risk from 60% to 5%, they were more likely to prefer the rest risk option over the zero-risk option (which was not the case when the rest-risk option was reduced from 25% to 5%). However, the bias was still found in this set of choices, especially in the social context, leading to the third finding: the bias is situational. Domain-specific risk-taking (Blais and Weber 2006; MacCrimmon and Wehrung 1990)—e.g., being risk-seeking in the recreational context and absolutely risk-averse in financial decision-making—can also be applied to the zero-risk bias. In our study (Schneider et al. 2017), participants were more biased (i.e., acted less rationally) when they had to make a social decision like moving with their company (and away from friends and/or family) than when they made investment or health decisions. In a social context, gaining perceived safety seems to be of high importance, which might be driven by the basic human need to belong (e.g., Baumeister and Leary 1995).

We propose to consider the bias as a simple heuristic ("A bird in the hand is worth two in the bush") which may be applied in several contexts. Heuristics are useful to facilitate decision-making in difficult or complex decision situations or when someone has incomplete knowledge. Gigerenzer and Gaissmaier (2011), for example, might categorize the mentioned problems as one-reason decisions which subsume several heuristics where only one cue is used in order to make a decision. The zero might serve as such a "clever cue" (Gigerenzer and Gaissmaier 2011). It might simply feel better and safer to have at least something—e.g., the bird in the hand—or to be able to exclude at least some kind of risk than to still live in uncertainty of gaining "two birds" (or to live in the fear of one more risk).

In economic choices it seems reasonable that people choose a free (often less valuable) product over a more valuable product they have to pay for. However, our own studies suggest that the zero-risk bias is especially strong in social contexts (Schneider et al. 2017). Thus, the effect cannot be explained by economic arguments only. One explanation why zero is such a special category for people is based on affective reactions. It has been argued that zero-risk options evoke more positive affect compared to remaining risk options and, as will be shown in the following section, affect can underlie decision-making (Loewenstein et al. 2001; Shampanier et al. 2007).

## 5.3 The Feeling of Safety

Often people rely on their feelings rather than statistical analyses when they judge risks in their daily lives. Feelings that arise in situations of risk are sensitive to the possibility of outcomes rather than the probability. While increases in probabilities do not generally map onto an increase in the emotional response, the small change from 0% to 1% does make a great difference for people, as the previous section has demonstrated. Especially when the consequences associated with a risk carry strong affective meanings (e.g., loss of lives), the probability of the event occurring often carries too little weight (Loewenstein et al. 2001; Rottenstreich and Hsee 2001). For example, people highly fear terrorist attacks, which are rather unlikely in comparison to other risks of dying such as heart attacks. People may focus on the outcome to the extent that they neglect probabilities. As a result, they fixate on extremely small risks with strong emotional responses (e.g., terrorism), rather than worry about larger risks that they face regularly (e.g., traffic accidents) (Sunstein 2003).

However, most everyday decisions under risk—such as crossing a busy street—do not provide probabilities for different options. Many scholars have therefore termed these as decisions under uncertainty rather than decisions under risk (e.g., Gigerenzer et al. 1999). In most cases, humans are not able to carefully analyze each possible outcome of uncertain decisions due to limitations in their cognitive capacities (cf. Simon, 1955). These limitations increase when quick decisions have to be made under time pressure, such as when driving a car (Finucane et al. 2000). In such a situation, we intuitively know when to brake or swerve even though we have not calculated the time it will take to approach the car in front of us. From an evolutionary perspective, this ability helped humans to survive, for example, by allowing us to quickly decide between fight or flight. Thus, quick reactions based on intuitive feelings are able to redirect cognitive processes towards potential sources of danger (Simon 1967).

When it comes to making decisions under risk and uncertainty, feelings play an important role. According to the *feelings-as-information hypothesis* (Schwarz and Clore 1983), subjective experiences (i.e., moods, emotions, bodily sensations) generally have an immediate influence on judgments. As a result, feelings that occur in a certain situation become a source of information. For example, if I feel positive around a person, I conclude that I like this person. However, feelings may lead one astray when they are unrelated to the situation and one is not aware of this unrelated source (e.g., bad mood due to bad weather).

Feelings or affective reactions usually precede cognitive analyses, which are based on objective information (Zajonc 1980). People might believe that they have analyzed all the options, went through all the pros and cons, and made a good decision based on the given facts, but often they already *feel* the answer all along. On the other hand, many people are aware that they have a gut feeling about the best option but are not able or willing to use this as a justification. For example, the experienced manager might just know that it is important to build a new factory, but spends hours with the team to come up with facts that support what he felt all along (Hoffrage and Marewski 2015). These examples highlight that feelings are an important source of information when making decisions or judging situations.

The link between risk (the counterpart of safety) and feelings has been intensively studied in psychology since the beginning of this century. Until then, decision-making under risk was mostly understood as a cognitive process and feelings received little attention. Slovic and colleagues (e.g., Slovic 2010; Slovic and Peters 2006; Slovic et al. 2003; Slovic et al. 2005) as well as Loewenstein and colleagues (e.g., Loewenstein et al., 2001; Loewenstein and Lerner 2003) integrated feelings in their research on decision-making under risk. Neuroscientists Bechara and Damasio (2005) stressed that "Emerging neuroscience evidence suggests that sound and rational decision making, in fact, depends on prior accurate emotional processing" (p. 336).

Slovic and colleagues have investigated people's perception of hazardous technologies and found that people's judgments of risks have little to do with possible outcomes and their actual probabilities. Their judgment is rather based on affective responses to the situation that are based on perceived dread (which includes perceived lack of control and catastrophic potential) and the risk of the unknown (Slovic et al. 2004). Dread and uncertainty may cause fear, which is a strong emotional reaction (Lerner and Keltner 2000). However, affective responses are more subtle feelings of "goodness" or "badness." To put it simply, a "bad feeling" makes most people feel unsafe. Reliance on affect in situations of risk and uncertainty has become known as the *affect heuristic* (Slovic 2010; Slovic and Peters 2006). Because people are not willing or not able to consider all information available, they simplify the decision-process by using a heuristic (e.g., Raue and Scholl 2018). Responses based on affect occur rapidly and automatically with or without conscious awareness.

Dual-process theories of thinking suggest that people process information in two fundamental ways. System 1 is intuitive, automatic, fast and experiential, while System 2 is analytical, deliberative, slow and rational. System 1 is associated with the experience of affect, which has also been described as the base of System 1 and explains why reliance on affect is quicker and easier than cognitive analysis (Epstein 1994; Evans 2003; Slovic et al. 2004; Stanovich and West 2000). However, feelings may be *biased* if the risk under consideration received recent media coverage, has a dramatic negative outcome, or was personally experienced, thus leading people to deviate from scientific data or reasoning (Tversky and Kahneman 1974). A study has shown, for example, that experienced forensic psychologists who were told that "20 out of every 100 patients similar to [this patient] are estimated to commit an act of violence" judged a patient as more dangerous than those who were presented with statistical information (i.e., a 20% chance of committing an act of violence). The authors of the study assumed that the relative frequency format (20 out of 100) created affect-laden images of violent patients, which was not the case with the percentage format (Slovic et al. 2000).

One of the major findings in the risk perception literature refers to an inverse relationship between perceived risk and perceived benefit. While risk and benefit are usually positively related (i.e., highly risky technologies are used because they are highly beneficial at the same time), *perceived* risk declines with an increase in *perceived* benefit and vice versa, resulting in a negative correlation of risk and benefit in people's minds. Subsequent studies found that this relationship can be explained by the affect heuristic: positive feelings result in a judgement of lower risks and higher benefits, while negative feelings result in a judgment of higher risk and lower benefits (Finucane et al. 2000; Slovic and Peters 2006). As a result, providing information on risk versus benefit can influence people's perceptions. For instance, this was shown for the perception of new technologies such as nuclear power (Finucane et al. 2000). These studies further support the affect heuristic by demonstrating that the inverse relationship between perceived risks and benefits increases under time pressure—a typical situation that activates System 1. In sum, studies have repeatedly shown that "affect influences judgment directly and is not simply a response to a prior analytic evaluation" (Slovic and Peters 2006, p. 323).

In research on the affect heuristic, affect has mainly an informational role that complements decision-making and offers alternative ways of action (Loewenstein et al. 2001). The risk-as-feelings hypothesis, on the other hand, states that emotions can also produce behavioral consequences that diverge from cognitive evaluations and do not align with the best course of action. For example, people who fear flying or public speaking may hinder their careers if these fears prevent them from traveling or giving presentations. Research has also shown that patients who lacked emotional reactions due to neurological impairment continued steering calmly when hitting ice on the road; healthy people, however, were more likely to slam their brakes in panic, causing the car to skid out of control (Shiv et al. 2005). These examples also highlight the divergence between cognitive evaluations and emotional reactions, as most people objectively recognize that getting in an airplane crash is highly unlikely and public speaking quite safe. Risk and safety perceptions, however, are often determined by emotional reactions rather than cognitive evaluations. Learning about the safety of airplanes does not change the behavioral reaction of an anxious flyer. On the other hand, experiencing fear in objectively risky situations has advantages. For example, patients with neurological impairments who lacked emotional reactions also engaged in much riskier investment choices because they did not experience emotional signals that would prevent them from taking high financial risks (Shiv et al. 2005).

The research discussed so far on feelings and decision-making under risk has mostly looked at integral emotions, which are feelings that are directly linked to the decision at hand. Incidental emotions, on the other hand, are feelings that are present but not relevant for the decision at hand (e.g., bad mood due to bad weather). However, incidental emotions may carry over from one situation to another and affect decision-making (see Lerner et al. 2015 for an overview). Early studies on these carry-over effects have shown that people who are in a bad mood make higher-risk judgments than people who are in a good mood even if their mood is unrelated to the judged risks (Johnson and Tversky 1983). In addition to the valence of emotions (i.e., positive vs. negative), Lerner and Keltner (2000, 2001) follow a multidimensional approach in their Appraisal-Tendency Framework (ATF). Their research has shown, for example, that fear increases risk judgment, while anger decreases risk judgments. Fear and anger are both emotions of negative valence, but they differ on other dimensions. Anger is a state of high certainty and control, while fear is a state of low certainty and control. As a result, negative events are perceived as more predictable and under human control in a state of anger rather than in a state of fear.

People's reliance on their feelings when judging risks can have consequences for public policy, when the public's perception of a risk diverges from the objective evaluation of experts. New policies might fail not because the cost-benefit analysis was poorly executed, but because the public has intense emotions created by extremely vivid images of negative outcomes (Sunstein 2003). Emotional reactions to a potential threat strongly increase when the threat can be easily imagined. The media is well-aware of this fact and uses strong images and personal stories that intensify people's emotional reactions and draws their attention to the story at hand. To create a feeling of safety among the public is extremely difficult when people are alarmed by their fears. At the same time, there are cases when the public pays too little attention to objectively dangerous threats such as climate change.

Several strategies have been suggested to minimize the unwanted influence of emotions in decision-making (Lerner et al. 2015). These include delaying the decision, which is often not feasible, and suppression of the emotion, which may backfire and intensify the emotion instead. More promising strategies are reappraisal (i.e., reframing the meaning of an emotion) or the induction of a counteracting emotional state (i.e., gratitude instead of sadness). A more recent approach that has gained much popularity in public policy is choice architecture and the use of nudges (Thaler and Sunstein 2008). Choice architects structure choices in a way that makes it easy for people to behave in their best interest without eliminating options. For example, choice architects may decide to set organ donation as a default with the possibility to opt out, which is a nudge—a little push in a good direction. Emotions such as fear may hinder the decision to become an organ donor and override cognitive evaluations of pros and cons. Mandatory waiting periods can also be nudges. For example, 21 States in the US require a waiting period of at least 24 hours between receiving the marriage license and the wedding. However, critics argue that choice architects still overlook emotions, which may limit the effectiveness of nudges (Lerner et al. 2015). Another approach that addresses people's feelings comes from the field of medical decision-making. As people struggle to make sense of probabilities (e.g., 10%), researchers have suggested using frequency formats (e.g., 10 in 100) to indicate the likelihood of side effects or the effectiveness of screenings in a format that is more easily imaginable (see Hoffrage and Garcia-Retamero 2018 for an overview). However, vivid experiences of family members suffering from a rare side effect or benefit from a screening may still override these more intuitive presentations of risks and skew people's perceptions of risks and/or benefits.

Every day we make several quick and automatic judgments around risks and safety using the experiential System 1, which draws from past experiences, images and stories that have feelings attached (Slovic and Peters 2006). This happens when we drive a car, cross a street or taste bad-smelling food. It is impossible to analyze these risks in such a short amount of time, and thus our feelings guide our decisions on what is safe. Fire fighters, for example, constantly have to make quick decisions in dangerous situations. Research has shown that firefighters do not make conscious choices or consider alternatives, but rather base their action on experience and constantly adapt it to the situation (Klein et al. 1986). However, experience can also have negative consequences in situations of risk, as the following section on learned carelessness will outline.

## 5.4 The Theory of Learned Carelessness

Frey and Schulz-Hardt's (1997) theory of learned carelessness provides a theoretical framework to explain risky—sometimes hazardous—behavior in everyday life (Frey et al. 2016). According to the theory, success (i.e., positive consequences of behavior) that has been reached with little effort due to luck or coincidence or repeatedly dangerous behavior without negative consequences results in a cognitive-affective monopoly hypothesis: "Everything is fine and will remain fine." This in turn influences perception (see hypothesis-based social perception; Lilli and Frey 1993). This monopoly hypothesis is built through learning experiences, namely positive (success without effort) and negative reinforcement (see also operant conditioning; Skinner 1963). Once learned carelessness is established, it can guide a person's perception, selection, and interpretation of subsequent information in favor of the monopoly hypothesis, which eventually may lead to risky behavior (Frey et al. 2016).

Carelessness is usually experienced as a positive state and a source of motivation. As a result, humans are susceptible to acting careless and forgoing precautions in favor of short-term positive feelings. Most people behave carelessly in many situations of their daily lives. For example, think about your own behavior: Do you wear a helmet when cycling to work? Do you secure your data on an external memory source? What about your password security? Did you recently eat junk food? Even though we *know* how to act wiser, we often still do not do it. In order to understand why, consider the following questions for those who cycle to work: (1) Have you succeeded (i.e., have you arrived at your workplace)? (2) Did you encounter any negative consequences (i.e., an accident)? (3) How did others (re)act (i.e., do they wear helmets)? (4) Did you enjoy your behavior or was it in some way less costly than the alternatives (i.e., you avoided ruining your hair)? Now, if you think about the next day, would you do it again?

What we colloquially described by our questions are the central aspects of the theory of learned carelessness. Carelessness can be seen as a "failure to give attention to, and take precautions against, the risks inherent in the successful prosecution of some activity" (White 1961, p. 593). Although most people know about the potential dangers of a certain behavior, many engage in it without taking reasonable precautions. By experiencing success without effort or through the repeated lack of negative consequences, people form and reinforce their perception that "everything is fine and will remain fine" (Frey and Schulz-Hardt 1997). Figure 5.2 illustrates the major assumptions and moderating factors of the theory.

There are two mediating factors which influence the formation and reinforcement of the monopoly hypothesis: social processes such as observational learning, norms and values as well as hedonism. In addition to individual learning experiences (positive and negative reinforcement), one can learn by observing the behavior of others and its consequences (Bandura 1977). If one observes that other people succeed with careless behavior, one will be more likely to show the same behavior. Furthermore, the more similar these other people are to oneself, the

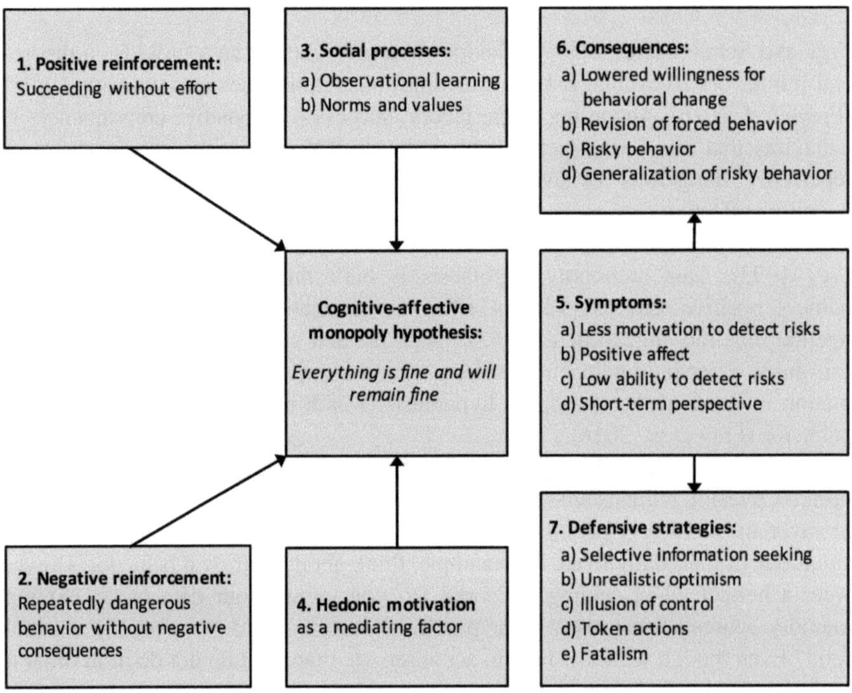


**Fig. 5.2** Simplified representation of the theoretical framework of learned carelessness (based on Frey and Schulz-Hardt 1997)

likelier it is that one will adopt the behavior (see similarity hypothesis in social comparison theory; Festinger 1954). Social norms and values are additional drivers of carelessness. If the social environment rewards risky or careless behavior, one is more likely to show such behavior in order to gain prestige or other incentives or to belong to the group. For example, social norms play a central role in careless behavior at the workplace. Mullen (2004) showed that workers abandoned protective equipment in order to portray a positive image in front of their colleagues. Regarding road safety, group pressure and risk-promoting norms (that can be transmitted by media) can lead to more careless behavior, resulting in increasing infringements and accidents (Shepherd et al. 2011).

The second mediating factor, hedonic motivation, refers to the "classic motivational principle that people approach pleasure and avoid pain" (Higgins 2006, p. 440). A high hedonic motivation fosters careless behavior as individuals neglect potential dangers in order to gain short-term pleasure. This is linked to delay discounting, a tendency to favor immediate but small rewards instead of larger, delayed rewards (Madden and Johnson 2010). Regarding risky behavior, research suggests that delay discounting is related to impulsivity and loss of control, which in turn are linked to lower academic performance, drug dependency and self-endangering behavior in general (e.g., Bickel and Marsch 2001; Kirby et al. 2005; Mischel et al. 1989).

Learned carelessness may result in motivational, affective, and cognitive symptoms. People may be less motivated to detect warning signals or potential dangers, as people generally tend to seek information that is congruent with their assumptions about the world (confirmation bias; e.g., Jonas et al. 2001). Research has demonstrated that people are more sensitive to information that signals positive consequences for the self than for information that signals negative consequences. Moreover, the lower the motivation to challenge one's own behavior, the lower the attention to and the processing of hazardous information (Frey 1986). In addition, both the assumption that "everything is fine and will be fine" and hedonic experiences from careless behavior lead to positive affective reactions. Positive feelings, in turn, are linked to lower risk perception and more risk-seeking behavior (Alhakami and Slovic 1994; Johnson and Tversky 1983; Slovic and Peters 2006).

Since carelessness often involves positive affect and promises pleasant states, people are more willing to search for and process information that justifies their behavior (Frey and Schulz-Hardt 1997). As a result, motivational and affective symptoms hinder thorough information-seeking and processing. Processing information that is congruent—that supports one's own view—is less cognitively effortful than processing incongruent information (e.g., Betsch and Glöckner 2010). Since cognitive capacity is limited, effortful information processing will only take place when there is enough capacity available. Time pressure, distracting cues, high informational load, or contradictory information consume cognitive capacity and may limit the processing of warning cues or reflection about one's own careless behavior (e.g., Betsch and Glöckner 2010; Gilovich et al. 2002; Hilbig et al. 2010; Kruglanski and Gigerenzer 2011).

In addition to these constraints in information processing, focusing on the short-term pleasures of risky or careless behavior and ignoring long-term negative effects is one of the symptoms of learned carelessness. Temporal distance plays a crucial role in risk perception: the closer in time the risky event, the more likelier it seems. On the other hand, if a risk might occur in the distant future, it is perceived as less likely and less threatening (Lermer et al. 2016; Raue et al. 2015; Trope and Liberman 2010; Wakslak and Trope 2009). Similar to delay discounting, immediate hedonic pleasures are overemphasized whereas associated long-term risks are underestimated.

The consequences of these symptoms are a lower willingness for behavioral change, the revision of forced behavior, risky behavior, and/or a potential generalization of risky behavior to other similar situations. Consider, for example, the prohibition of using cell phones without hands-free equipment while driving. At first, talking on the phone while driving was not a problem. Due to an increase in accidents caused by people using a cell phone without hands-free equipment, talking on the phone while driving was prohibited by law. However, once a behavior is established, it is hard to change and people are unwilling to abandon the behavior. If people are forced to abandon their habit, they will still find opportunities to use their phones while driving (revision of forced behavior). In addition, if they cannot do it in the car, they could still do it on the bike (generalization to similar situations).

Learned carelessness may also result in several defensive strategies to maintain the monopoly hypothesis and justify one's careless behavior (Frey and Schulz-Hardt 1997). These defensive strategies can be seen on a continuum of coping with a perceived threat, beginning at the very low level of selective information seeking (1) all the way to culminating in fatalism (5) (see Fig. 5.3).

*Selective information* seeking stands at the very beginning of this continuum. People ignore, deny, or suppress information that indicates risks and dangers of their own careless behavior. By simply ignoring the risks of smoking for one's health, one can still enjoy a cigarette. The next step is *unrealistic optimism*. One knows about the dangers but is convinced that it will not happen to herself. Both strategies are used to maintain perceived safety and to avoid change. Feeling able to control the environment or influencing the outcomes by one's own actions (*illusion of control*) is an important factor for creating perceived safety. If someone perceives a high controllability of dangerous or risky events, he will perceive less threat and

| Perceived Threat | | | | |
|---|---|---|---|---|
| 1. Selective information seeking | 2. Unrealistic optimism | 3. Illusion of control | 4. Token actions | 5. Fatalism |
| Hazard information is ignored, suppressed or neglected | Perception and acknowledgement of a general danger, but not a personal danger | Acknowledgment of being personally in danger but belief to be in control of the situation | Token actions in order to silence one's conscience | Maintaining the behavior because no alternative actions or ways to influence the situation are seen |

**Fig. 5.3** Defensive strategies to maintain learned carelessness (Frey et al. 2016)

he will be less likely to take precautions. A smoker can reduce a perceived threat by assuming that he can quit smoking tomorrow and convince himself that he just does it for taste. The fourth defensive strategy refers to *token actions*. These are actions that are less effortful than a behavioral change and have the benefit of silencing one's conscience. Instead of quitting smoking, one might start exercising two times a week for health reasons. *Fatalism* is at the endpoint of this continuum. If one does not perceive any control or if the threat is too high, one might give up hope and take a fatalistic view: "Never mind! You only live once!"

By integrating various psychological theories into one theoretical framework, the theory of learned carelessness helps to understand and explain needlessly risky behavior. Through learning processes, the monopoly hypothesis that "everything is fine and will remain fine" is formed, which in turn can be reinforced by social norms and values, observational learning, and hedonic motives. Its symptoms are an uncritically sublime mood, superficial information-seeking and processing as well as a focus on short-term rewards without considering long-term detriments. These factors promote unnecessary risky behavior, which can be frequently observed in day-to-day life. Practical implications can be drawn by considering and eliminating defensive mechanisms that justify carelessness and hinder behavioral change.

## 5.5 Summary and Practical Implications

In this chapter, we offered some insights into situational influences on perceived safety. Due to their inability to consciously consider all the information available in a situation, humans simplify judgments and use heuristics. Reducing a risk from 10% to zero makes a decision easier than having to consider a risk reduction from 25% to 5%. This preference for zero risk may not always be in people's best interest, but the perception of safety seems to dominate this decision-making strategy. As was outlined in the chapter, cognitive evaluations of decision-making tasks are often overridden by feelings present in the situations. People strive for positive emotional experiences and want to avoid negative emotional experiences. Thus, the affect heuristic is another decision-making strategy that can be useful to simplify a task. However, feelings may lead one astray when they are unrelated to the task at hand or when people experience strong emotional reactions such as fear or panic. As a result, the subjective feeling of safety may differ from the objective recognition (i.e., the cognitive evaluation) of safety and even from the evaluation of experts (Loewenstein et al. 2001). Feelings are a product of the vivid imagination of consequences, personal exposure or past experiences, just to name a few examples. But having developed a strong feeling of safety may also backfire, as has been acknowledged by the theory of learned carelessness. Repeated learning experiences of behaving riskily without negative consequences can result in an underestimation of risk.

What can be done to combat these biases? Risk communication plays a crucial role in the perception of safety. As people's risk perceptions often deviate from objective probabilities, risks need to be communicated in a way that embraces the way humans process information. People are especially susceptible to the biasing factors mentioned in this chapter (i.e., overweighing certainty, being guided by feelings or positive learning experiences) when sufficient information is not available or too complex to process. This happens, for example, when news outlets put horrible images of a terrorist attack on their front pages, which immediately increases people's perceptions of the likelihood of such an attack (e.g., Slovic et al. 2005). The field of medical decision-making has proposed several ways of presenting risk information in ways that support transparency, understanding and assessment of health-relevant information. For example, presenting screening results (i.e., likelihood of breast cancer after receiving a positive finding in a mammography screening) in natural frequencies (e.g., 990 out of 1000 women receive a negative result, but 9 out of these 990 actually do have breast cancer) rather than probabilities (e.g., 1% of breast cancers are overlooked in screenings).[1] Women hope to feel safer and have a 'peace of mind' after going to a mammography screening and receiving a negative result, but even doctors are often not able to draw the right conclusions from given probabilities (Eddy 1982; Hoffrage and Garcia-Retamero 2018; Hoffrage et al. 2000). Visual aids may increase the understanding of health-relevant information even more and support intuitive processing (Garcia-Retamero and Hoffrage 2013). Hoffrage and Garcia-Retamero (2018) point out that "a negative [mammogram] result cannot be equated with security," but presenting this information in natural frequencies can support patients (and also doctors) in weighing the benefits and costs of a screening. In addition, people's perceptions of risk and safety can be dramatically influenced by presenting risk in relative versus absolute numbers (Yamagishi 1997). Assuming that four women out of 200 would die without screening, one of these four could be saved through the screening, which results in a relative risk reduction through the screening of 25%. However, one could also state that 3 instead of 4 out of 200 women would die, which results in an absolute risk reduction of 0.5% (Hoffrage and Garcia-Retamero 2018).

In general, people strive for safety and zero risk, but real life usually does not offer that. Ensuring transparency and intuitive understanding of risks is an important step to support decision-making in situations of uncertainty. But what happens if people are not motivated to engage in cognitive considerations? We have learned that emotions and learning experiences can bias judgments—often unconsciously. Among the most influential intervention strategies to combat these unconscious biases are nudges (Thaler and Sunstein 2008). Nudges are used by choice architects to structure a choice in a way that makes it easier for people to make better decisions and minimize biases. Typical nudges are defaults or

[1]See Nelson et al. (2016) for a recent study on rates of false-positive and false-negative results from digital mammography screening.

automatic enrollments (e.g., Benartzi and Thaler 2013; Johnson and Goldstein 2003), re-framing of information (Cox and Cox 2001; Gifford and Comeau 2011) or leveraging people's tendency to conform to social norms by providing comparisons with peers (e.g., Allcott 2011; Raue et al. 2018). However, nudges have been criticized for their underlying assumption that people are not able to decide in their best interest. These critics suggest that people should instead be supported in their decision-making competency (e.g., Gigerenzer 2015; Schlag 2010; Wilkinson 2013).

On a final note, we should recognize our amazing ability to make effective decisions in complex situations despite limited cognitive capabilities. The heuristics we use do not always result in errors and biases; in fact, they serve us well in many decisions we make. *Take the zero option! Go for what feels good! It didn't harm me in the past!* are heuristics that simplify decisions and result in good judgments in many situations. The trick is to realize when it is time to get support—from System 2 or from choice architects—to ensure safety.

**Acknowledgements** We thank Adam Felts for proofreading this chapter.

## References

Ajzen, I. (1991). The theory of planned behavior. *Organizational Behavior and Human Decision Processes, 50*(2), 179–211.

Alhakami, A. S., & Slovic, P. (1994). A psychological study of the inverse relationship between perceived risk and perceived benefit. *Risk Analysis, 14*(6), 1085–1096.

Allais, M. (1953). Le comportement de l'homme rationnel devant le risque: Critique des postulats et axiomes de l'ecole Americaine [The rational individual's behavior in the presence of risk: Critique of the postulates and axioms of the American school]. *Econometrica, 21,* 503–546. https://doi.org/10.2307/1907921.

Allcott, H. (2011). Social norms and energy conservation. *Journal of Public Economics, 95*(9–10), 1082–1095.

Atkinson, J. W. (1957). Motivational determinants of risk-taking behavior. *Psychological Review, 64*(6), 359–372.

Bandura, A. (1977). *Social learning theory*. New York, NY: General Learning Press.

Baron, J., Gowda, R., & Kunreuther, H. (1993). Attitudes toward managing hazardous waste: What should be cleaned up and who should pay for It? *Risk Analysis, 13*(2), 183–192.

Baumeister, R. F., & Leary, M. R. (1995). The need to belong: Desire for interpersonal attachments as a fundamental human motivation. *Psychological Review, 117*(3), 497–529.

Bechara, A., & Damasio, A. R. (2005). The somatic marker hypothesis: A neural theory of economic decision. *Games and Economic Behavior, 52*(2), 336–372.

Benartzi, S., & Thaler, R. H. (2013). Behavioral economics and the retirement savings crisis. *Science, 339,* 1152–1153.

Betsch, T., & Glöckner, A. (2010). Intuition in judgment and decision making: Extensive thinking without effort. *Psychological Inquiry, 21*(4), 279–294.

Bickel, W. K., & Marsch, L. A. (2001). Toward a behavioral economic understanding of drug dependence: delay discounting processes. *Addiction, 96,* 73–86.

Blais, A. R., & Weber, E. U. (2006). A Domain-Specific Risk-Taking (DOSPERT) scale for adult populations. *Judgment and Decision Making, 15*(4), 263–290.

Chauvin, B. (2018). Individual differences in the judgment of risks: Sociodemographic characteristics, cultural orientation, and level of expertise. In M. Raue, E. Lermer, & B. Streicher (Eds.), *Psychological perspectives on risk and risk analysis—Theory, models and applications* (pp. 37–61). New York: Springer.

Cox, D., & Cox, A. D. (2001). Communicating the consequences of early detection: The role of evidence and framing. *Journal of Marketing, 65,* 91–103.

Eddy, D. M. (1982). Probabilistic reasoning in clinical medicine: Problems and opportunities. In D. Kahneman, P. Slovic, & A. Tversky (Eds.), *Judgment under uncertainty* (pp. 249–267). Cambridge: Cambridge University Press.

Eller, E., Streicher, B., & Lermer, E. (2012). Psychologie und Risikomanagement: Warum wir Risiken falsch einschätzen [Psychology and risk management: Why we judge risks wrongly]. *Risiko Manager, 23,* 6–9.

Epstein, S. (1994). Integration of the cognitive and the psychodynamic unconscious. *American Psychologist, 49*(8), 709–724.

Evans, J. S. B. T. (2003). In two minds: Dual-process accounts of reasoning. *Trends in Cognitive Sciences, 7*(10), 454–459.

Festinger, L. (1954). A theory of social comparison processes. *Human Relations, 7,* 117–140.

Finucane, M. L., Alhakami, A., Slovic, P., & Johnson, S. M. (2000). The affect heuristic in judgments of risks and benefits. *Journal of Behavioral Decision Making, 13,* 1–17.

Fishbein, M., & Ajzen, I. (1975). *Belief, attitude, intention and behavior: An introduction to theory and research.* Reading: Addison-Wesley.

Frey, D. (1986). Recent research on selective exposure to information. *Advances in Experimental Social Psychology, 19,* 41–80.

Frey, D., & Schulz-Hardt, S. (1997). Eine Theorie der gelernten Sorglosigkeit [A theory of learned carelessness]. In *Bericht über den 40. Kongress der Deutschen Gesellschaft für Psychologie* (pp. 604–611).

Frey, D., Ulrich, B., Streicher, B., Schneider, E., & Lermer, E. (2016). Theorie der gelernten Sorglosigkeit [Theory of learned carelessness]. In H.-W. Bierhoff & D. Frey (Eds.), *Enzyklopädie der Psychologie. Sozialpsychologie [Encyclopedia of Psychology]* (pp. 429–469). Göttingen, Germany.

Garcia-Retamero, R., & Hoffrage, U. (2013). Visual representation of statistical information improves diagnostic inferences in doctors and their patients. *Social Science and Medicine, 83,* 27–33.

Gifford, R., & Comeau, L. A. (2011). Message framing influences perceived climate change competence, engagement, and behavioral intentions. *Global Environmental Change, 21*(4), 1301–1307.

Gigerenzer, G. (2015). On the supposed evidence for libertarian paternalism. *Review of Philosophy and Psychology, 6*(3), 361–383.

Gigerenzer, G., & Gaissmaier, W. (2011). Heuristic decision making. *Annual Review of Psychology, 62*(1), 451–482.

Gigerenzer, G., Todd, P. M., & The ABC Research Group. (1999). Simple heuristics that make us smart. New York: Oxford University Press.

Gilovich, T., Griffin, D., & Kahneman, D. (2002). *Heuristics and biases: The psychology of intuitive judgment.* Cambridge: Cambridge University Press.

Hart, S. (2005). Adaptive heuristics. *Econometrica, 73*(5), 1401–1430.

Hastie, R., & Dawes, R. M. (2001). A normative, rational decision theory. In R. Hastie & R. M. Dawes (Eds.), *Rational choice in an uncertain world. The psychology of judgement and decision making* (pp. 249–287). Thousand Oaks: Sage.

Higgins, E. T. (2006). Value from hedonic experience and engagement. *Psychological Review, 113*(3), 439–460.

Hilbig, B. E., Scholl, S. G., & Pohl, R. F. (2010). Think or blink—Is the recognition heuristic an "intuitive" strategy? *Judgment and Decision Making, 5*(4), 300–309.

Hoffrage, U., & Garcia-Retamero, R. (2018). Improving Understanding of health-relevant numerical information. In M. Raue, E. Lermer, & B. Streicher (Eds.), *Psychological perspectives on risk and risk analysis—Theory, models and applications* (pp. 279–298). New York: Springer.

Hoffrage, U., & Marewski, J. N. (2015). Unveiling the lady in black: Modeling and aiding intuition. *Journal of Applied Research in Memory and Cognition, 4*(3), 145–163.

Hoffrage, U., Lindsey, S., Hertwig, R., & Gigerenzer, G. (2000). Communicating statistical information. *Science, 290*(5500), 2261–2262.

Johnson, E. J., & Goldstein, D. (2003). Do defaults save lives? *Science, 302*(5649), 1338–1339.

Johnson, E. J., & Tversky, A. (1983). Affect, generalization, and the perception of risk. *Journal of Personality and Social Psychology, 45*(1), 20–31.

Jonas, E., Schulz-Hardt, S., Frey, D., & Thelen, N. (2001). Confirmation bias in sequential information search after preliminary decisions: An expansion of dissonance theoretical research on selective exposure to information. *Journal of Personality and Social Psychology, 80*(4), 557.

Kahneman, D., & Tversky, A. (1979). Prospect theory: An analysis of decision under risk. *Econometrica Journal of the Econometric Society, 47,* 263–292.

Kirby, K. N., Winston, G. C., & Santiesteban, M. (2005). Impatience and grades: Delay-discount rates correlate negatively with college GPA. *Learning and Individual Differences, 15*(3), 213–222.

Klein, G. A., Calderwood, R., & Clinton-Cirocco, A. (1986). Rapid decision making on the fire ground. *Proceedings of the Human Factors and Ergonomics Society Annual Meeting, 30*(6), 576–580.

Kruglanski, A. W., & Gigerenzer, G. (2011). Intuitive and deliberate judgments are based on common principles. *Psychological Review, 118*(1), 97–109.

Lauriola, M., & Weller, J. (2018). Beyond daredevils: Risk taking from a temperament perspective. In M. Raue, E. Lermer, & B. Streicher (Eds.), *Psychological perspectives on risk and risk analysis—Theory, models and applications* (pp. 3–369). New York: Springer.

Lermer, E., Streicher, B., Sachs, R., Raue, M., & Frey, D. (2016). Thinking concretely increases the perceived likelihood of risks: The effect of construal level on risk estimation. *Risk Analysis*, 623–637.

Lerner, J. S., & Keltner, D. (2000). Beyond valence: Toward a model of emotion-specific influences on judgement and choice. *Cognition and Emotion, 14*(4), 473–493.

Lerner, J. S., & Keltner, D. (2001). Fear, anger, and risk. *Journal of Personality and Social Psychology, 81*(1), 146–159.

Lerner, J. S., Li, Y., Valdesolo, P., & Kassam, K. S. (2015). Emotion and decision making. *Annual Review of Psychology, 66*(1), 799–823.

Lilli, W., & Frey, D. (1993). Die Hypothesentheorie der sozialen Wahrnehmung [Theory of hypothesis-based social perception]. In D. Frey & M. Irle (Eds.), *Theorien der Sozialpsychologie, Band 1: Kognitive Theorien [Theories in social psychology, volume 1: Cognitive theories]* (2nd ed., pp. 49–80). Bern: Huber.

Loewenstein, G. F., Weber, E. U., Hsee, C. K., & Welch, N. (2001). Risk as feelings. *Psychological Bulletin, 127*(2), 267–286.

Loewenstein, G., & Lerner, J. S. (2003). The role of affect in decision making. In R. Davidson, H. Goldsmith, & K. Scherer (Eds.), *Handbook of affective science* (pp. 619–642). Oxford: Oxford University Press.

MacCrimmon, K. R., & Wehrung, D. A. (1990). Characteristics of risk taking executives. *Management Science, 36*(4), 422–435.

Madden, G. J., & Johnson, P. S. (2010). A delay-discounting primer. In G. J. Madden & W. K. Bickel (Eds.), *Impulsivity: The behavioral and neurological science of discounting* (pp. 11–37). Washington, DC: American Psychological Association.

Mather, M., Mazar, N., Gorlick, M. A., Lighthall, N. R., Burgeno, J., Schoeke, A., et al. (2012). Risk preferences and aging: The "certainty effect" in older adults' decision making. *Psychology and Aging, 27*(4), 801–816.

Mischel, W., Shoda, Y., & Rodriguez, M. I. (1989). Delay of gratification in children. *Science, 244*(4907), 933–938.

Mullen, J. (2004). Investigating factors that influence individual safety behavior at work. *Journal of Safety Research, 35*(3), 275–285.

Nelson, H. D., O'Meara, E. S., Kerlikowske, K., Balch, S., & Miglioretti, D. (2016). Factors associated with rates of false-positive and false-negative results from digital mammography screening: An analysis of registry data. *Annals of Internal Medicine, 164*(4), 226–235.

Raue, M., & Scholl, S. G. (2018). The use of heuristics in decision-making under risk and uncertainty. In M. Raue, E. Lermer, & B. Streicher (Eds.), *Psychological perspectives on risk and risk analysis—Theory, models and applications* (pp. 153–179). New York, NY: Springer.

Raue, M., D'Ambrosio, L. A., & Coughlin, J. F. (2018). The power of peers: Prompting savings behavior through social comparison. *Manuscript submitted for publication.*

Raue, M., Streicher, B., Lermer, E., & Frey, D. (2015). How far does it feel? Construal level and decisions under risk. *Journal of Applied Research in Memory and Cognition, 4*(3), 256–264.

Ritov, I., Baron, J., & Hershey, J. C. (1993). Framing effects in the evaluation of multiple risk reduction. *Journal of Risk and Uncertainty, 6*(2), 145–159.

Rottenstreich, Y., & Hsee, C. K. (2001). Money, kisses, and electric shocks: On the affective psychology of risk. *Psychological Science, 12*(3), 185–190.

Schlag, P. (2010). Nudge, Choice Architecture, and Libertarian Paternalism. *Michigan Law Review, 108*(6), 913–924.

Schneider, E., Streicher, B., Lermer, E., Sachs, R., & Frey, D. (2017). Measuring the zero-risk bias: Methodological artefact or decision-making strategy? *Zeitschrift Für Psychologie, 225*(1), 31–44.

Schwarz, N., & Clore, G. L. (1983). Mood, misattribution, and judgments of well-being: Informative and directive functions of affective states. *Journal of Personality and Social Psychology, 45*(3), 513–523.

Shampanier, K., Mazar, N., & Ariely, D. (2007). Zero as a special price: The true value of free products. *Marketing Science, 26*(6), 742–757.

Shepherd, J. L., Lane, D. J., Tapscott, R. L., & Gentile, D. A. (2011). Susceptible to social influence: Risky "driving" in response to peer pressure. *Journal of Applied Social Psychology, 41*(4), 773–797.

Shiv, B., Loewenstein, G., Bechara, A., Damasio, H., & Damasio, A. R. (2005). Investment behavior and the negative side of emotion. *Psychological Science, 16*(6), 435–439.

Simon, H. A. (1955). A behavioral model of rational choice. *The Quarterly Journal of Economics, 69*(1), 99–118.

Simon, H. A. (1967). Motivational and emotional controls of cognition. *Psychological Review, 74*(1), 29–39.

Skinner, B. F. (1963). Operant behavior. *Psychological Review, 18*(8), 503–515.

Slovic, P. (2010). *The feeling of risk: New perspectives on risk perception*. London: Routledge.

Slovic, P., & Peters, E. (2006). Risk perception and affect. *Current Directions in Psychological Science, 15*(6), 322–325.

Slovic, P., Finucane, M. L., Peters, E., & MacGregor, D. G. (2004). Risk as analysis and risk as feelings: Some thoughts about affect, reason, risk, and rationality. *Risk Analysis, 24*(2), 311–322.

Slovic, P., Finucane, M., Peters, E., & MacGregor, D. G. (2003). The affect heuristic. In T. Gilovich, D. Griffin, & D. Kahneman (Eds.), *Heuristics and biases: The psychology of intuitive judgment* (pp. 1–51). Cambridge: Cambridge University Press.

Slovic, P., Monahan, J., & MacGregor, D. G. (2000). Violence risk assessment and risk communication: The effects of using actual cases, providing instruction, and employing probability versus frequency formats. *Law and Human Behavior, 24*(3), 271.

Slovic, P., Peters, E., Finucane, M. L., & MacGregor, D. G. (2005). Affect, risk, and decision making. *Health Psychology, 24*(4S), S35.

Stanovich, K. E., & West, R. F. (2000). Advancing the rationality debate. *Behavioral and Brain Sciences, 23*(05), 701–717.

Statistisches Bundesamt. (2015). Gestorbene durch Unfälle. Retrieved June 12, 2018, from https://www.destatis.de/DE/ZahlenFakten/GesellschaftStaat/Gesundheit/Todesursachen/Tabellen/Sterbefaelle_Unfaelle.html.

Sunstein, C. R. (2003). Terrorism and probability neglect. *Journal of Risk and Uncertainty, 26*(2), 121–136.

Svenson, O. (1981). Are we all less risky and more skillful than our fellow drivers? *Acta Psychologica, 47*(2), 143–148.

Thaler, R. H., & Sunstein, C. R. (2008). *Nudge*. New York: Penguin.

Trope, Y., & Liberman, N. (2010). Construal-level theory of psychological distance. *Psychological Review, 117*(2), 440–463.

Tversky, A., & Kahneman, D. (1974). Judgment under uncertainty: Heuristics and biases. *Science, 185,* 1124–1131.

Tversky, A., & Kahneman, D. (1992). Advances in prospect theory: Cumulative representation of uncertainty. *Journal of Risk and Uncertainty, 5*(4), 297–323.

Van den Bos, K. (2009). Making sense of life: The existential self trying to deal with personal uncertainty. *Psychological Inquiry, 20*(4), 197–217.

van den Bos, K., Poortvliet, P. M., Maas, M., Miedema, J., & van den Ham, E.-J. (2005). An enquiry concerning the principles of cultural norms and values: The impact of uncertainty and mortality salience on reactions to violations and bolstering of cultural worldviews. *Journal of Experimental Social Psychology, 41*(2), 91–113.

Viscusi, W. K. (1997). Alarmist decisions with divergent risk information. *The Economic Journal, 107*(445), 1657–1670.

Viscusi, W. K., Magat, W. A., & Huber, J. (1987). An investigation of the rationality of consumer valuations of multiple health risks. *The Rand Journal of Economics, 18*(4), 465.

Visschers, V. H. M., & Siegrist, M. (2018). Differences in risk perception between hazards and between individuals. In M. Raue, E. Lermer, & B. Streicher (Eds.), *Psychological perspectives on risk and risk analysis—Theory, models and applications* (pp. 63–80). New York: Springer.

Wakslak, C., & Trope, Y. (2009). The effect of construal level on subjective probability estimates. *Psychological Science, 20*(1), 52–58.

Weinstein, N. D. (1989). Optimistic biases about personal risks. *Science, 246*(4935), 1232–1234.

White, A. R. (1961). Carelessness, indifference and recklessness. *The Modern Law Review, 24*(5), 592–595.

Wilkinson, T. M. (2013). Nudging and manipulation. *Political Studies, 61*(2), 341–355.

Yamagishi, K. (1997). When a 12.86% mortality is more dangerous than 24.14%: Implications for risk communication. *Applied Cognitive Psychology, 11*(6), 495–506.

Zajonc, R. B. (1980). Feeling and thinking: Preferences need no inferences. *American Psychologist, 35*(2), 151–175.

# Chapter 6
# The Assessment of Risk Perception: Influence of Answer Format, Risk Perspective and Unrealistic Optimism

**Eva Lermer, Bernhard Streicher, Martina Raue and Dieter Frey**

**Abstract** Risk is a very complex construct and, as such, can be studied from various perspectives. In the present chapter we discuss the influences of answer format on risk assessment, the chosen perspective from which a risk is assessed and describe two systems of probabilistic reasoning. Although many different measurements exist today, there is no clear methodological advice as to which kind of instrument is appropriate in a given context. As a suggestion, we present the risk assessment matrix (RAM), a model developed to explore the impact of both answer format and risk perspective on risk assessment. The RAM combines both factors and addresses the question of how the measurement of subjective perceived risk is influenced by psychological factors triggered by methodological aspects.

**Keywords** Risk assessment · Answer format · Scales · Probabilistic reasoning · Unrealistic optimism · Risk assessment matrix

E. Lermer (✉)
Department of Experimental Psychology, University of Regensburg,
Universitätsstr. 31, 93053 Regensburg, Germany
e-mail: eva.lermer@ur.de

E. Lermer
FOM University of Applied Sciences for Economics and Management,
Arnulfstr. 30, 80335 Munich, Germany

B. Streicher
Institute of Psychology, UMIT—Private University for Health Sciences,
Medical Informatics and Technology, Eduard-Wallnöfer-Zentrum 1,
6060 Hall in Tyrol, Austria
e-mail: bernhard.streicher@umit.at

M. Raue
AgeLab, Massachusetts Institute of Technology, 77 Massachusetts Avenue,
Cambridge, MA 02139, USA
e-mail: maraue@mit.edu

D. Frey
Department Psychology, Ludwig-Maximilian-University Munich, Leopoldtr. 13,
80802 Munich, Germany
e-mail: dieter.frey@psy.lmu.de



M. Raue et al. (eds.), *Perceived Safety*, Risk Engineering,
https://doi.org/10.1007/978-3-030-11456-5_6

## 6.1 Introduction

The handling of risks is becoming more and more important for economy, science and society. Therefore, it is not without reason that Adams (2001) assumes safety, the counterpart to risk, to be the world's largest industry. However, risk is a very complex construct and, as such, can be studied from various perspectives. This might be the major reason why research on measuring risk perceptions appears to be a precious collection of single findings, despite a relatively long tradition of research in this field (especially since the works of Tversky and Kahneman 1974; Slovic 1987; Weinstein 1999). On the one hand, risk perceptions are influenced by many different factors like context, domain, traits, states and severity. On the other hand, risk assessments are strongly subject to biases, which are, up to now, not well understood (Price 2001). All in all, previous research on this topic has two key shortcomings: the methodological approach of risk assessment, and the understanding of the underlying psychological processes. In this chapter, we outline the impact of answer format, risk perspective and unrealistic optimism on risk assessment. Furthermore, we present a theoretical model (called risk assessment matrix), which considers these influences and is useful for researchers and practitioners when assessing risk estimates.

## 6.2 The Influence of Answer Format on Risk Assessment

It is very likely that there is no golden trail towards a single answer to the question of how to measure risk perceptions. To date, it is still not clear which answer format should be used for measuring quantitative risk assessments. A review of the relevant literature reveals that there are few studies dealing with the evaluation of different question formats in risk assessments. For example, risk assessment was explored with different numerical linear scales (Diefenbach et al. 1993; Weinstein et al. 2007; Woloshin et al. 2000), 1 in *x* scales (Cuite et al. 2008; Woloshin et al. 2000), logarithmic scales (Diefenbach et al. 1993; Woloshin et al. 2000), percentage scales (Cuite et al. 2008; Diefenbach et al. 1993; Slovic et al. 2000; Weinstein et al. 2007; Yamagishi 1997) and frequency scales (Cuite et al. 2008; Slovic et al. 2000; Yamagishi 1997; please see Fig. 6.1). In summary, there is no clear advice as to when which scale should be used for assessing risk perceptions.

The above list presents only some of the numerous common formats that are possible. However, with the exception of verbal scales (e.g., from *1* = *no chance* to *7* = *certain*), there is not much agreement on when which scale should be used (Lermer et al. 2013). Research on verbal scales has shown that this kind of measurement has some advantages due to its intuitive nature, in particular, when compared to numerical scales. Verbal scales correspond to people's everyday way of thinking, i.e. how people actually think about probabilities (e.g. "I think x is likely"; see also Raue and Schneider, Chapter 5). Windschitl and Wells (1996)

*Verbal and numerical linear scales*

| | | | | | | |
|---|---|---|---|---|---|---|
| No chance | Very unlikely | Unlikely | Moderate chance | Likely | Very likely | Certain to happen |

No chance 1 2 3 4 5 6 7 Certain to happen

-5 -4 -3 -2 -1 0 1 2 3 4 5

*Closed 11-step percentage scale*

| | | | | | | | | | | |
|---|---|---|---|---|---|---|---|---|---|---|
| [ ] | [ ] | [ ] | [ ] | [ ] | [ ] | [ ] | [ ] | [ ] | [ ] | [ ] |
| 0% | 10% | 20% | 30% | 40% | 50% | 60% | 70% | 80% | 90% | 100% |

*Open percentage scale*: ___ %

*Closed (logarithmic) frequency scale*

| | | | | | | | | |
|---|---|---|---|---|---|---|---|---|
| 1 | 2 | 3 | 4 | 5 | 6 | 7 | 8 | 9 |
| 1:1 | 1:10 | 1:100 | 1:1.000 | 1:10.000 | 1:100.000 | 1:1.000.000 | 1:10.000.000 | 1:100.000.000 |

*Semi open frequency scale*: **1 out of *n***

*Open frequency scale*: ***X* out of *Y***

*Visual analog scale*

No chance |——————————————| certain

**Mixed forms:**

*Verbally and numerically anchored linear scales*

| | | | | | | | | |
|---|---|---|---|---|---|---|---|---|
| No chance | 1 | 2 | 3 | 4 | 5 | 6 | 7 | certain |
| | | Probably will not happen | | 50/50 chance | | Probably will happen | | |

*Logarithmic frequency and percentage scale*

| | | | | | | | | |
|---|---|---|---|---|---|---|---|---|
| [ ] | [ ] | [ ] | [ ] | [ ] | [ ] | [ ] | [ ] | [ ] |
| No chance | | 1 out of 1.000 (0.1%) | | 1 out of 100 (1%) | | 1 out of 10 (10%) | | Certain (100%) |

**Fig. 6.1** Commonly used answer formats (adapted from Lermer et al. 2016a)

revealed that numeric measures have the tendency to prompt a more deliberate assessment of probability judgments as opposed to verbal scales. As a result, the widely used numerical measures do not necessarily reflect the way most people assess probabilities or think about uncertainties in the real world. People mostly tend to think intuitively and not deliberately (Kahneman 2012). This is why verbal scales have a higher usability (also because they are easy to comprehend;

Diefenbach et al. 1993). In other words, people are more used to expressing their probability judgments verbally and not numerically; for example, the majority of individuals answer the question, "How high do you estimate the risk that it will rain tomorrow?", with words like "not very likely" or "perhaps". Only a few people would respond with numerical answers such as: "I think there is a 30% chance of rain tomorrow". This also means that verbal scales are better suited to predicting future behavior (for instance vaccination, Weinstein et al. 2007). However, this raises the problem of objectivity, and the inter- and intrapersonal comparability of peoples' answers, which is not given with verbal scales. A "perhaps" can mean a higher probability for one person than for someone else. However, numerical (quantitative) answer formats are also not straight forward. Numerical formats can be complex in three ways:

### *6.2.1 Numeric Scales Require a Certain Level of Numerical Competence*

In general, differentiations in the range of 0 and 1% are difficult for most people to grasp: Without appropriate practice, people generally do not immediately understand such a very low probability of—for example—1:100,000 when expressed as a percentage value of 0.001%. Lermer et al. (2013) have shown that risk estimates for very rare events, which were assessed with an *open percentage scale,* were considerably higher than the same estimates measured with an *open* or *closed frequency scale*. Similar results were also found by Slovic, Monahan and McGregor (Slovic et al. 2000). The reason for this effect is presumed to lie in the different representation of absolute and relative frequencies. People are more familiar with absolute frequencies than with thinking in probabilities. Furthermore, people are strongly impressed by absolute frequencies. For example, in a study by Yamagishi (1997), participants were presented with the number of deaths due to well-known causes. In this study, people judged the riskiness of death due to for example cancer higher when described as "kills 1,286 out of 10,000" as when described as "kills 24.14 out of 100 (Yamagishi 1997). The different perception of absolute and relative frequencies is particularly relevant for risk communication (see for instance, Hoffrage et al. 2000; Hoffrage and Gigerenzer 1998; Lermer et al. 2018).

### *6.2.2 Numerical Format Triggers a Need for Accuracy*

People strive to make accurate statements, regardless of whether an intuitive or deliberate kind of thinking is activated. However, our need to give accurate answers is particularly prevalent when we think deliberately about certain issues. To explain this aspect, let us digress briefly to consider some psychological research on intuitive vs. deliberate thinking. Kahneman (2012) has shown that people use two different kinds of thinking: the intuitive way, which he calls system 1 and a deliberate way, which he calls system 2. System 1 operates quickly and automat ically whereas system 2 requires much more cognitive attention. For instance, when we are asked the question "how much is 2 + 2?" we can effortlessly give the right answer, thanks to our system 1. However, when the problem is more complex (e.g., "17 × 89 = ?"), we must concentrate to perform well and use System 2. Generally, system 1 is our default system. System 2 starts to work when it has to (e.g., when we are confronted with a problem that cannot be automatically solved) and switches off as soon as possible. Furthermore, our intuition (system 1) can be very strong and override our system 2. Let us use an example: "A bat and a ball cost $1.10. The bat costs one dollar more than the ball. How much does the ball cost?" (Kahneman 2012, p. 44; Frederick 2005). Most people intuitively answer 10¢, which is not correct (the correct answer is 5¢). Therefore, we sometimes need to resist our intuition in order to end up with the correct answer.

This is particularly relevant when assessing perceived risk. Generally, answering via a numerical scale activates deliberate thinking while answering via a verbal scale activates intuitive thinking. This is due to the inherent characteristic of the response format. When answering a verbal scale (e.g., ranging from *1 = very unlikely* to *7 = very likely*), we easily find arguments for our answer when we have to justify our decision (e.g., why we have responded with "unlikely" and not "very unlikely"). This, however, is not the case when the answer format is, for instance, a percentage scale. It is also worth noting that answers to numerical scales can be compared with actual statistics. For risk estimates, it is likely that there is one correct answer from which our answer may deviate a little or a lot. This determines the quality of our estimates. Thus, numerical scales trigger a need for exactness. Therefore, it is important to note that numerical formats can lead to results that do not exactly reflect how people actually think about risks (because such answers tend to be more intuitive; for an overview see Kahneman 2012). For instance, Windschitl and Wells (1996) and Lermer et al. (2013) showed that verbal scales are considerably more sensitive to psychological effects (like *unrealistic optimism*, see below) than numerical formats. However, in addition to the kind of answer format, different ranges for the same answer format can also influence risk estimates.

### 6.2.3 The Range of the Scale Influences the Result

Studies have shown that quantitative judgments are strongly influenced by the given ranges of the answer formats (e.g., Schwarz 1999; Schwarz et al. 1991). For instance, Schwarz et al. (1991) compared the impact of linear scales; one scale ranged from *−5* to *+5* and the other from *0* to *10*. Results showed that using the former scale led to significantly higher estimates than the latter (for questions like: "How successful have you been in life, so far?", Schwarz et al. 1991, p. 572). The authors point to the negative numbers and the zero point within the scale as an

**Table 6.1** The influence of answer formats on risk assessment (adapted from Slovic 2010, p. 103)

| Version 1 | Version 2 |
|---|---|
| No more likely | No more likely |
| Twice as likely | Twice as likely |
| 5 times as likely | 3 times as likely |
| 10 to 20 times as likely | 5 times as likely |
| 50 times as likely | 10 or more times as likely |

explanation: it is much more unpleasant to rate oneself with a negative value than with a positive one (for further criticism of scale formats with a negative range, see Harris and Hahn 2011, who also highlight the influence of scale format on unrealistic optimism).

Another example of the influence of the range for the answer scale on risk assessment includes an US telephone survey of about 3,500 participants conducted by the Annenberg Public Policy Center of the University of Pennsylvania (2000). Current smokers were asked (amongst other things), "If you smoked a pack of cigarettes a day, how much do you think it would increase your chances of getting lung cancer?" (Slovic 2010, p. 103). Two answer formats were provided (Table 6.1).

Results showed that the ranges in the response format had a strong impact on participants' estimates. In *version 1,* 41.1% of adult smokers estimated an increase of 10 times or more (total of both answer alternatives: *10 to 20 times as likely* and *50 times as likely*), whereas only 14.6% of adult smokers choose this range in version 2 (i.e., *10 or more times as likely*). The difference was also marked in the answers of younger participants (aged 14 to 22): 57.1% in version 1 to 30.9% in version 2. Thus, results showed that the level of risk assessment was decisively influenced by the range of the answer formats.

Thus, it is important to consider the following when assessing perceived risk: (1) the numerical skills of the respondents, (2) whether intuitive or conscious reasoning is in play, and (3) the range of the answer format. However, there are some further important psychological influences on the assessment of perceived risk, which can arise due to the chosen risk perspective.

## 6.3 Unrealistic Optimism

Another important aspect of risk assessment is the chosen perspective from which a risk is assessed. Research has shown that it makes a great difference in risk estimates whether it is oneself who could be affected by a negative event or someone else (e.g., average person). Generally, people assume that their own future is more positive than the fate of others (Lermer et al. 2013, 2016b). One generally accepted explanation of why perspective biases the perception of risk is *unrealistic optimism* (Harris and Hahn 2011; Klein and Helweg-Larsen 2002; McKenna 1993; Regan et al. 1995; Weinstein 1980, 1989). For *it won't happen to me* assumption, most studies assume a downward comparison (Wills 1981). This means that a comparison of oneself with (real or imagined) less fortunate others increases one's own subjective well-being when negative outcomes are expected (Harris and Hahn 2011; Perloff and Fetzer 1986).

A number of research studies have revealed the effect of downward comparison by asking subjects to estimate their perceived probability of being affected by a negative scenario for themselves and also for another target (e.g., an average person or a student; McKenna 1993; Robertson 1977; Weinstein 1980). Perloff and Fetzer (1986) were among the first to show that the level of abstraction of the target person (e.g., average person vs. closest friend) plays a decisive role. Their research showed that participants who were asked to estimate risks (e.g., heart attack) for themselves, for specific targets (e.g., their closest friend) and for abstract persons (e.g., average student or *one of your friends*) make downward comparisons as soon as they have the opportunity to do so. In other words, people's estimates of their own personal risk and the risk of another person depends on who the other person is and how he/she is presented. For instance, in one case study by Perloff und Fetzer (Perloff and Fetzer 1986), when participants were asked to estimate their personal risk and the risk of *one of their friends* (a very vague description), they tended to choose (as the comparison target) someone out of their friends who was more vulnerable to negative events (compared to themselves). Here, personal risk estimates were significantly lower than those for the comparison target. Thus, people tend to show unrealistic optimism when they have the possibility to select a comparison target. However, participants who were forced to think about a specific close person did not estimate their risk as lower than that of, say for example, their closest friend as the comparison target. These results indicate that participants perceived no difference between their own vulnerability and that of the specific person.

Thus, the comparison target description (vague/abstract vs. specific) is decisive for the estimated level of risk. When the comparison target is abstract, people have the tendency to be unrealistically optimistic about their own personal risk. In other words, when we are asked to compare our risk with the risk of someone else, we prefer a comparison person who is less fortunate. However, another probable reason for the discrepancy between personal risk estimates and those for abstract targets, is explained by the following theoretical considerations regarding probabilistic reasoning.

## 6.4 Probabilistic Reasoning

Reeves and Lockhart (1993) showed that people use two different systems for probabilistic reasoning when estimating likelihoods: a distributional and a singular approach. If a person is described vaguely (e.g., average person), then that person is perceived as one unit of a class. For example, if we want to estimate the risk of a person getting divorced, we most likely think about the target person as a "married person". As a result, the target person becomes one vague/abstract unit of the class "married persons". Then, in order to estimate the probability for divorce, we apply a distributional approach und derive our judgments from a perceived base rate (i.e., the general frequency of divorce in a population). However, if the person is perceived as a unique and specific entity, and attention is only given to the individual (e.g., person X), a singular approach will be used. Here, probability estimates are based on the evaluation of this specific person's dispositions and attitudes.

These two probabilistic systems (cf. Kirkpatrick and Epstein 1992; Gigerenzer 1991; Kahneman and Tversky 1979) are particularly relevant to risk assessment. They enable the link from an inferential strategy (distributional or singular approach) to the context of a judgment (Reeves and Lockhart 1993). People often seem to equate available base-rate information with the risk probability for an abstract person (e.g., equating a person's risk of divorce with the perceived base-rate for divorce; Klar et al. 1996). However, when the person becomes specific, people neglect base-rate information and derive their estimates from attributed variables. For instance, the estimated risk that person X's marriage will end in divorce depends on what is attributed to X. Accordingly, a distributional approach is used when people assess risks of a *person*, which is a very abstract term. This, in general, leads to higher risk estimates compared to a singular approach, which is used when people assess their personal risks or risks of a *specific other* (i.e., named person). Thus, for risk assessment it is very important how abstract the target person is described.

Up to this point, we have shown that risk estimates are influenced by different factors. We presented various common answer formats for assessing perceived risk and described the influence of the chosen answer format (e.g., numeric vs. verbal) and format range on risk estimates. Furthermore, we highlighted the importance of the risk perspective, especially in regard to the effect of unrealistic optimism and also the used terms (e.g., abstract vs. specific person). People have the tendency to be unrealistically optimistic when they compare their personal risk with the risk of an abstract target person (but not/less so when the comparison target is specific). Moreover, people seem to use different systems of probabilistic reasoning for different targets (abstract vs. specific). In the literature, these influences are inseparable. Therefore, we propose an integrated approach to risk assessment. This is important for a better understanding of how people perceive and estimate risks. As a consequence, we developed a theoretical risk assessment matrix (RAM) model, which combines the influences of answer format, risk perspective and unrealistic optimism on risk assessment. The RAM has been empirically tested by Lermer et al. (2013) and is described below.

## 6.5 The Risk Assessment Matrix

Overall, it remains unclear which scale and response format is appropriate for a particular situation, and which psychological mechanisms explain the ascertained differences in response behavior between scales. It may well be, that less is known about psychological biases in risk estimations than was previously assumed. In order for the understanding of risk assessment to be applied, further exploration of factors influencing risk assessment is necessary. Lermer et al. (2013) developed a model called risk assessment matrix (RAM) to explore the impact of both answer format and risk perspective on risk assessment. The RAM combines both factors and addresses the question of how the measurement of subjective perceived risk is influenced by psychological factors triggered by methodological aspects.

In particular, the authors explored the influence of four different questionnaire answer formats (*rating scale*, *open percentage scale*, *open* and *closed frequency scale*) and of four different risk perspectives (*abstract person*, *self*, *specific person*, *specific others*). Thus, the study comprised 16 different questionnaire versions (see Fig. 6.2).

All participants were asked to estimate 20 risk probabilities (e.g., airplane crash or serious car accident), which were the same in all 16 conditions. Only the perspective (i.e., target person to which the risks were related) and the response format varied. This means, in the condition *person* only the term *person* was given (without any further information). Participants assigned to this perspective, were asked to estimate the 20 risk probabilities either on a *rating scale*, an *open frequency scale*, an *open percentage scale*, or on a *closed frequency scale*. The instruction text reads: "How high do you estimate the risk of a person to be affected by one of the following events during his/her lifetime?" (Lermer et al. 2013). This pattern was the same for the remaining three perspectives. In the condition *self*, participants were asked to estimate their own personal risks, and in the condition *specific other,* the term *person* was changed to *Anton* or *Petra*. Thus, by naming the abstract target, the target became specific. However, no other information (except the name) was provided. The condition *specific others* varied. For each of the 20 risk items, a different male or female name was added. Here the instruction text

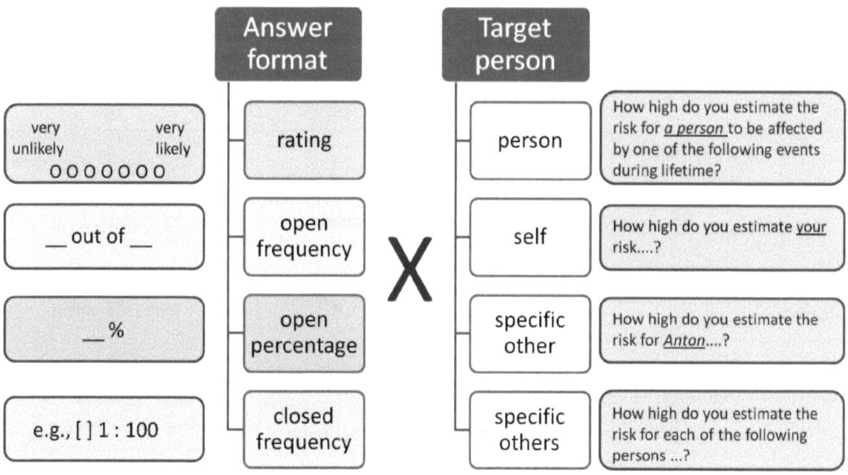


**Fig. 6.2** Study design (Lermer 2013, p. 38)

read: "How high do you estimate the risk of each of the following persons to be affected by the following events during his/her lifetime?" (e.g., "Fin is robbed"; Lermer et al. 2013).

The idea of the RAM is to systematically consider the influence of both the answer format and risk perspective. As discussed above, research has shown that quantitative answer formats (e.g., *percentage scale*) tend to trigger a more deliberate kind of thinking, whereas qualitative answer formats (e.g., *rating scale*) leave more freedom for intuitive answers. This differentiation is displayed along rows (i.e., answer format: intuitive vs. deliberative) in the RAM (see Fig. 6.3). The kind of probabilistic reasoning due to the risk perspective is represented as columns (probabilistic reasoning: distributional vs. singular).

Overall, results showed that risk estimates for an abstract target were significantly higher than those for a specific target (e.g., *self* or *Anton*). Like Reeves and Lockhardt (1993), Lermer et al. (2013) attribute this difference to the influence of base-rates. Thus, considering the risk of an abstract target leads to higher risk estimates because it is not a specific individual that comes to mind but a base-rate. However, if the person in question is not considered a high-risk person, it is very likely that equating the base-rate with the person's risk probability leads to an overestimation of the abstract person's risk. The left column in the RAM states that when a distributional approach is used, people tend to consider base-rates, irrespective of the chosen answer format (intuitive or deliberate). The right column shows the opposite: when people think of the risk associated with a specific person, they tend not to consider base-rates and use a singular approach in their probabilistic reasoning. Thus, it is very likely that these risk estimates are smaller than those when people have base-rates in mind.

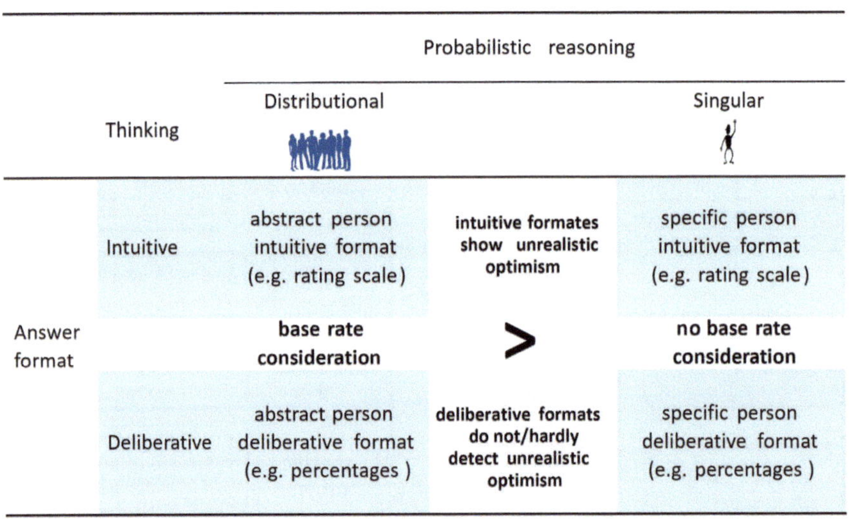


**Fig. 6.3** Risk Assessment Matrix (RAM; Lermer 2013, p. 38)

Another important aspect of the RAM is that it considers *unrealistic optimism* within these different ways of assessing subjective perceived risk. Carrying on from previous research, Lermer et al. (2013) showed that risk estimates for one's self were significantly lower than those for abstract targets. Furthermore, when the comparison target was specific (i.e., *Anton/Petra*), risk estimates for abstract targets (i.e., *a person*) were higher. However, not every answer format displayed these effects: results for intuitive answer formats (e.g., rating scale) quite clearly showed unrealistic optimism effects (i.e., higher risk estimates for abstract target) whereas this was not always the case for deliberate formats (e.g. percentages scale). People tend to show less unrealistic optimism when they are confronted with answer formats requiring more cognitive effort. These effects are displayed in the middle column of the RAM.

In summary, the RAM provides some useful tips for risk assessment, that involve both the choice of answer format and risk perspective (prior to assessing perceived risks), and support for the interpretation of already collected data on risk assessments. We further discuss some practical implications of how the RAM and other approaches may contribute to a valid assessment of perceived risk.

## 6.6 Practical Implications

Risk estimates are assessed and interpreted in many different scientific disciplines (e.g., psychology, economics and medicine) and business contexts (e.g., insurance and banking industry, consulting firms). Although many different measurements exist today, there is no clear methodological advice as to which kind of instrument is appropriate in a given context. Going back to an earlier step, in most cases, there is not even an awareness of the diverging consequences arising from the use of different instruments used to assess subjective perceived risk. We highlight the importance of a reasoned choice of answer format and risk perspective because the way risks are assessed has a tremendous impact on the outcome. Thus, the following questions should be asked before subjective risk estimates are assessed:

(A) *What is the aim of the assessment?*

Is it to find out people's intuitive thoughts about the risk probabilities? There is much to suggest that intuitive thoughts are most valid for assessing people's everyday way of thinking. Or is it to get to know the results of deliberate probabilistic reasoning of the perception of the level of risk, in order to get to the most accurate outcome (which does not necessarily reflect the way people actually think about risks)?

In practical terms, verbal scales are advocated if people's actual and intuitive perceived risk estimates are sought. Verbal scales are intuitive, easy to comprehend, and reflect most people's way of thinking of probabilities. However, one should note that there are some downsides to using verbal scales: We do not know how people individually assess the given verbal labels (e.g., a "likely" may mean something different to different people). Therefore, interpersonal risk estimate comparisons should be handled with caution, and involve consideration of deviations due to subjective interpretations of the verbal terms. This is obvious in comparisons of intrapersonal risk estimates: When people, for instance, have to estimate two totally different risks, like “die from alcohol liver disease” vs. “die from flu”, it may be possible that they perceive both risks as “very unlikely”. However, when “very unlikely” is one end of a scale, we cannot differentiate between these two estimates, although they may be very different for the individual. Moreover, risk estimates assessed with verbal scales cannot be analyzed with currently available statistics. Therefore, it is almost impossible (without any numerical anchor values) to determine the accuracy of the risk estimates (except by solely ranking comparisons).

This brings us to a second point, numerical scales are recommended when the intention is to investigate people’s deliberate probabilistic reasoning and their accuracy in assessing risks. Numerical scales allow for inter- and intrapersonal comparisons as well as comparison with existing statistics. The downsides are that numerically assessed risk estimates do not necessarily reflect the way people think about risks in their everyday life (where they generally use an intuitive kind of thinking) and, also, many people have tremendous difficulties with numerical formats (especially percentages). This leads to the next question, which should be raised before assessing perceived risk.

(B) *How experienced are the respondents with answering numerical, quantitative answer formats, especially very low probabilities?*

In general, people are not used to thinking about probabilities in a numerical way (rather more verbally; e.g., something is *likely*). Moreover, very low probabilities (e.g., a negative event happening to one person out of 10 million) are very hard to image adequately without having previous experience in this kind of thinking. This could be one reason for inconsistent results when risk estimates are compared interpersonally.

In practical terms, it is, therefore, advisable to find out the numerical skills of the persons questioned in advance of them assessing risks in a numerical way, especially when the scenarios are very unlikely. If the participants are not very numerically skilled, one should weigh the pros and cons of a numerical answer format and, if necessary, forego a quantitative assessment in favor of a valid investigation. Furthermore, in choosing a numerical assessment, the answer format should be adjusted to the specific risk probability context. This means that if, for instance, very low probabilities (e.g., especially between 0 and 1% chance of occurrence) are to be assessed, then one should consider scales that focus on this probability spectrum. Furthermore, it is important to note that frequency formats are more comprehensible than percentages. In other words, the context/issue should determine the answer format. This also applies to the risk perspective.

(C) *What risk perspective is taken?*

Is it really the personal risk perception that is of interest? There are many reasons to assume that personal risk estimates are biased downwards when compared to risk estimates for another person due to self-enhancement. Alternatively, do people predominantly have the base-rate (relative frequency of affected persons) in mind when assessing the respective risk?

In practical terms, one should consciously define the perspective when assessing perceived risk. There is a tremendous difference between whether one has to estimate their own personal risk or the risk of someone else. When it is the personal perceived risk that is of interest, the answer is clear. However, when the base-rate people have in mind is of interest, then it might be advisable to use abstract targets (e.g., "a person") in combination with a frequency value (e.g., "one person out of 100"), which is, in general, more comprehensible then percentages. In other words, one should keep in mind that the level of abstraction of the target person (i.e., labeling the potentially affected person) influences the probabilistic strategy and, in turn, the risk estimates.

Once risk estimates are collected, the RAM aids interpretation of the results; both the chosen answer format (intuitive vs. deliberative) and perspective (abstract vs. specific). However, more research is needed for a better understanding of the psychological processes involved when perceived risks are assessed. Hereto, it would be very interesting to take up the RAM-approach and to expand on it, particularly in regard to the influence of both answer format and target person on risk assessment across different risk domains (e.g., sports vs. finance vs. gambling vs. health) and different degrees of risk severity (e.g., the extent of losses: monetary value vs. number of victims). Moreover, it may reasonably be assumed that different cultural backgrounds (i.e., individualism vs. collectivism) may also lead to different patterns (e.g., see Hsee and Weber 1999; Weber and Hsee 1998; Weber et al. 1998).

## 6.7 Conclusion

In summary, the present work highlights different methodological influences on risk assessment that are relevant to the measurement of risk estimates. Research shows that subjective probability estimates on negative events are very sensitive. The chosen answer format, as well as the respective risk perspective, trigger different cognitive processes, which lead to deviating results. For instance, the combination of a *percentage scale* with the risk perspective *person* leads to estimates that are very high. This is probably due to the use of a distributional approach as well as the scale inherent character of the percentage format. In contrast, the use of a *closed frequency scale* with a specific person risk perspective (or personal risk perspective) leads to lower estimates. This is probably due to the singular approach, and in the case of personal risk estimates, due to self-enhancement leading to unrealistic optimism effects. The RAM describes which combination leads to higher/lower risk estimates due to methodological and some cognitive reasons. Thus, it is obvious that risk assessments are not only subject to subjective estimates. This might be one reason why there is not much common advice currently available on how to measure subjective risk estimates (Nicholson et al. 2005). However, this chapter aims to shed some light on what can bias risk assessment and people's perceptions of risks.

## References


Adams, J. (2001). *Risk*. London: Routledge.

Cuite, C. L., Weinstein, N. D., Emmons, K., & Colditz, G. (2008). A test of numeric formats for communicating risk probabilities. *Medical Decision Making, 28*(3), 377–384. https://doi.org/10.1177/0272989X08315246.

Diefenbach, M. A., Weinstein, N. D., & O'Reilly, J. (1993). Scales for assessing perceptions of health hazard susceptibility. *Health Education Research, 8*(2), 181–192. https://doi.org/10.1093/her/8.2.181.

Frederick, S. (2005). Cognitive reflection and decision making. *Journal of Economic Perspectives, 19*(4), 25–42.

Gigerenzer, G. (1991). How to make cognitive illusions disappear: beyond "heuristics and biases". *European Review of Social Psychology, 2,* 83–115. https://doi.org/10.1080/14792779143000033.

Harris, A. J. L., & Hahn, U. (2011). Unrealistic optimism about future life events: a cautionary note. *Psychological Review, 118*(1), 135–154. https://doi.org/10.1037/a0020997.

Hsee, C. K., & Weber, E. U. (1999). Cross-national differences in risk preference and lay predictions. *Journal of Behavioral Decision Making, 12,* 165–179. https://doi.org/10.1002/(SICI)1099-0771(199906)12:2%3c165:AID-BDM316%3e3.0.CO;2-N.

Hoffrage, U., & Gigerenzer, G. (1998). Using natural frequencies to improve diagnostic inferences. *Academic Medicine, 73*(5), 538–540.

Hoffrage, U., Lindsey, S., Hertwig, R., & Gigerenzer, G. (2000). Communicating statistical information. *Science, 290*(5500), 2261–2262.

Kahneman, D., & Tversky, A. (1979). Prospect theory: an analysis of decision under risk. *Econometrica: Journal of the Econometric Society, 47,* 263–292.

Kahneman, D. (2012). *Thinking fast and slow*. London: Allen Lane.

Kirkpatrick, L. A., & Epstein, S. (1992). Cognitive-experiential self-theory and subjective probability: further evidence for two conceptual systems. *Journal of Personality and Social Psychology, 63*(4), 534–544.

Klar, Y., Medding, A., & Sarel, D. (1996). Nonunique invulnerability: singular versus distributional probabilities and unrealistic optimism in comparative risk judgments. *Organizational Behavior and Human Decision Processes, 67*(2), 229–245.

Klein, C. T. F., & Helweg-Larsen, M. (2002). Perceived control and the optimistic bias: a meta-analytic review. *Psychology and Health, 17*(4), 437–446. https://doi.org/10.1080/0887044022000004920.

Lermer, E. (2013). The impact of answer format and construal level on risk assessment and behavior. Dissertation, Ludwig Maximilian University Munich, Germany.

Lermer, E., Raue, M., & Frey, D. (2016a). Risikowahrnehmung und Risikoverhalten [Risk perception and risk-taking behavior]. In D. Frey & H-W. Bierhoff (Hrsg.) Enzyklopädie der Psychologie—Sozialpsychologie (pp. 535–580). Göttingen: Hogrefe.

Lermer, E., Streicher, B., & Raue, M. (2018). Measuring subjective risk estimates. In M. Raue, B., E. Lermer, & B. Streicher (Eds.), *Psychological perspectives on risk and risk analysis: theory, models and applications*. New York: Springer.

Lermer, E., Streicher, B., Sachs, R., & Frey, D. (2013). How risky? The impact of target person and answer format on risk assessment. *Journal of Risk Research, 16*(7), 903–919. https://doi.org/10.1080/13669877.2012.761267.

Lermer, E., Streicher, B., Sachs, R., Raue, M., & Frey, D. (2016b). Thinking concretely increases the perceived likelihood of risks: The effect of construal level on risk estimation. Risk Analysis, 36, 623-637. https://doi.org/10.1111/risa.12445.

McKenna, F. P. (1993). It won´t happen to me: unrealistic optimism or illusion of control? *British Journal of Psychology, 84,* 39–50. https://doi.org/10.1111/j.2044-8295.1993.tb02461.x.

Nicholson, N., Soane, E., Fenton-O'Creevy, M., & Willman, P. (2005). Personality and domain-specific risk taking. *Journal of Risk Research, 8*(2), 157–176. https://doi.org/10.1080/1366987032000123856.

Perloff, L. S., & Fetzer, B. K. (1986). Self-other judgments and perceived vulnerability to victimization. *Journal of Personality and Social Psychology, 50,* 302–310. https://doi.org/10.1037//0022-3514.50.3.502.

Price, P. C. (2001). A group size effect on personal risk judgments: implications for unrealistic optimism. *Memory and Cognition, 29*(4), 578–586. https://doi.org/10.3758/BF03200459.

Reeves, T., & Lockhart, R. (1993). Distributional versus singular approaches to probability and errors in probabilistic reasoning. *Journal of Experimental Psychology: General, 122*(2), 207–226. https://doi.org/10.1037//0096-3445.122.2.207.

Regan, P. C., Snyder, M., & Kassin, S. M. (1995). Unrealistic optimism: self-enhancement or person positivity? *Personality and Social Psychology Bulletin, 21,* 1073–1082. https://doi.org/10.1177/01461672952110008.

Robertson, L. S. (1977). Car crashes: perceived vulnerability and willingness to pay for crash protection. *Journal of Community Health, 3,* 136–141. https://doi.org/10.1007/BF01674235.

Schwarz, N., Knäuper, B., Hippler, H., Nölle-Neumann, E., & Clark, L. (1991). Rating scales numeric values may change the meaning of scales. *Public Opinion Quaterly, 55,* 570–582. https://doi.org/10.1086/269282.

Schwarz, N. (1999). How the Questions Shape the Answers. *American Psychologist, 54*(2), 93–105. https://doi.org/10.1037//0003-066X.54.2.93.

Slovic, P. (1987). Perception of risk. *Science, 236,* 280–285. https://doi.org/10.1126/science.3563507.

Slovic, P. (2010). Cigarette smokers: Rational actors or rational fools? In P. Slovic (Ed.), *The feeling of risk: new perspectives on risk perception* (pp. 85–108). London: Earthscan Publications.

Slovic, P., Monahan, J., & MacGregor, D. (2000). Violence risk assessment and risk communication: the effects of using actual cases, providing instruction, and employing probability versus frequency formats. *Law and Human Behavior, 24*(3), 271–296. https://doi.org/10.1023/A:1005595519944.

Tversky, A., & Kahneman, D. (1974). Judgment under uncertainty: heuristics and biases. *Science, 185*(4157), 1124–1131. https://doi.org/10.1126/science.185.4157.1124.

Weber, E. U., & Hsee, C. K. (1998). Cross-cultural differences in risk perception, but cross-cultural similarities in attitude towards perceived risk. *Management Science, 44,* 1205–1217. https://doi.org/10.1287/mnsc.44.9.1205.

Weber, E. U., Hsee, C. K., & Sokolowska, J. (1998). What folklore tells us about risk and risk taking: a cross-cultural comparison of American, German, and Chinese proverbs. *Organizational Behavior and Human Decision Processes, 75,* 170–186. https://doi.org/10.1006/obhd.1998.2788.

Weinstein, N. D. (1980). Unrealistic optimism about future life events. *Journal of Personality and Social Psychology, 39,* 806–820. https://doi.org/10.1037//0022-3514.39.5.806.

Weinstein, N. D. (1989). Optimistic biases about personal risks. *Science, 246,* 1232–1233. https://doi.org/10.1126/science.2686031.

Weinstein, N. D. (1999). What does it mean to understand a risk? Evaluating risk comprehension. *Journal of the National Cancer Institute Monographs, 25,* 15–20. https://doi.org/10.1093/oxfordjournals.jncimonographs.a024192.

Weinstein, N. D., Kwitel, A., McCaul, K. D., Magnan, R. E., Gerrard, M., & Gibbons, F. X. (2007). Risk perceptions: assessment and relationship to influenza vaccination. *Health Psychology: Official Journal of the Division of Health Psychology American Psychological Association, 26*(2), 146–151. https://doi.org/10.1037/0278-6133.26.2.146.

Wills, T. A. (1981). Downward comparison principles in social psychology. *Psychological Bulletin, 90,* 245–271. https://doi.org/10.1037//0033-2909.90.2.245.

Windschitl, P., & Wells, G. L. (1996). Measuring psychological uncertainty: Verbal versus numeric methods. *Journal of Experimental Psychology: Applied, 2*(4), 343–364. https://doi.org/10.1037//1076-898X.2.4.343.

Woloshin, S., Schwartz, L. M., Byram, S., Fischhoff, B., & Welch, H. G. (2000). A new scale for assessing perceptions of chance: a validation study. *Medical Decision Making, 20*(3), 298–307. https://doi.org/10.1177/0272989X0002000306.

Yamagishi, K. (1997). When a 12.86% mortality is more dangerous than 24.14%: implications for risk communication. *Applied Cognitive Psychology, 11,* 495–506. https://doi.org/10.1002/(SICI)1099-0720(199712)11:6%3c495:AID-ACP481%3e3.0.CO;2-J.

# Part II
# Practical Examples of Perceived Safety

# Chapter 7
# The Concept of Risk Perception in Health-Related Behavior Theory and Behavior Change

**Susanne Gaube, Eva Lermer and Peter Fischer**

**Abstract** Messages aiming to increase the public's perception of health and safety risks, such as the spread of antibiotic-resistant pathogens, are omnipresent. In these cases, the basic assumption often is that a heightened level of risk perception should lead to more protective behaviours like proper hand hygiene in hospitals. The notion that people's perception of health risks influences their risk-taking or safety behaviour is prevalent both in health behaviour theories and applied health communication. However, research findings on the connection between risk perception and health-related behaviour are not clear-cut. In the present chapter, we look at the different operationalisations of the term risk perception and discuss several methodological issues that are widespread in the health risk perception literature which might have led to inconclusive results. Overall, even though the effect sizes are generally moderate, the majority of research findings indicate that risk perception influences health- and safety-related behaviour. This was shown both in research looking at a variety of different health-related behaviours at the same time as well as in studies only concerned with specific activities such as hand hygiene and vaccination. Therefore, risk perception as a concept truly deserves its place in health behaviour theory and behaviour change interventions. Some implications of these findings on intervention design are discussed.

**Keywords** Risk perception · Risk · Affect · Health-related behaviour · Hand hygiene · Vaccination

S. Gaube (✉) · E. Lermer · P. Fischer
Department of Experimental Psychology, University of Regensburg,
Universitätsstr. 31, 93053 Regensburg, Germany
e-mail: susanne.gaube@ur.de

E. Lermer
FOM University of Applied Sciences for Economics & Management,
Arnulfstr. 30, 80335 Munich, Germany
e-mail: eva.lermer@ur.de

P. Fischer
e-mail: peter.fischer@ur.de

 
M. Raue et al. (eds.), *Perceived Safety*, Risk Engineering,
https://doi.org/10.1007/978-3-030-11456-5_7

## 7.1 Introduction

*"Killer germs: Death from the hospital"* (von Leszczynski, 2016).

Somewhat counterintuitive to people's striving for safety and security, such exaggerated, fear-evoking headlines regarding health risks are prevalent in the mass media (Reynolds, 2001). Sensational accounts of health risks grab the audience's attention, which should encourage people to read or listen to the ensuing story (Hastings, Stead, & Webb, 2004). The article following the headline above was meant to inform readers about the risks regarding healthcare-associated infections, which pose a threat to the health and safety of millions of patients worldwide. The main cause for healthcare-associated infections is a lack of hygiene standards, including non-compliance toward hand hygiene guidelines. The problem is intensified by the rise of antibiotic-resistant pathogens (World Health Organisation [WHO], 2009). Newspaper articles like the above often include calls for relevant behaviour change at a personal or societal level—for example, to optimise hand hygiene behaviour in healthcare facilities and decrease the widespread use of antibiotics to reduce the infection risk of antibiotic-resistant pathogens. It is also a common approach in governmental or institutional health communication strategies aiming to increase public safety to emphasise the risks of engaging in certain behaviours (i.e., smoking increases the vulnerability for lung cancer) or the risk from neglecting safety behaviours (i.e., higher infection rates in clinics with low levels of hand hygiene) (Ruiter, Kessels, Peters, & Kok, 2014; Sheeran, Harris, & Epton, 2014). The underlying idea is that increasing people's risk perception should lead to lower levels of risk-taking and more protective and safety-enhancing behaviours (Sheeran et al., 2014). A simple illustration would be that if a physician's awareness of the risk of transmitting dangerous pathogens via her hands increases, she will practice better hand hygiene behaviour to protect herself and increase patient safety.

The basic concept that risk perception influences behaviour is persuasive in common sense and is an integral part of many major health behaviour theories. Health behaviour theories seek to describe, explain, predict and eventually help to modify human behaviour (Brewer, Chapman, Gibbons, Gerrard, McCaul, & Weinstein, 2007). For instance, risk perception is explicitly embedded in the *extended parallel process model* (Witte, 1992), the *health belief model* (Rosenstock, 1966), the *protection motivation theory* (Rogers, 1983), and the *health action process approach* (HAPA; Schwarzer, 2008), which are all well-researched and validated models of the genesis of health-related behaviour (Sheeran, Klein, & Rothman, 2017). To give one example, in the HAPA there is a distinction between a *pre-intentional motivational phase,* in which the behavioural intention is formed, and a *post-intentional volition phase* that leads up to the actual behaviour (Schwarzer, 2008). The determinants of intention are: (1) *risk perception* (e.g., "I am at risk of transmitting dangerous pathogens via my hands"), (2) *outcome expectations* (e.g., "If I clean my hands before patient contact, I will reduce my risk of transmitting pathogens"), and (3) *self-efficacy* (e.g., "I am capable of adhering to hand hygiene guidelines despite time constraints"). After the intention towards a health behaviour is formed, it has to be manifested through additional processes (*planning, resources and barriers, self-efficacy,* and *action control*) to lead to action. Within the HAPA, risk perception alone is not sufficient to form a person's behavioural intention nor explain actual health-related behaviour but paves the way for further cognitive decision processes (Schwarzer, Lippke, & Luszczynska, 2011). In other well-known behaviour theories, risk perception is more implicitly integrated as a determining factor for health-related behaviours (van der Pligt, 1996). For instance, Ajzen (1991) postulated in his *theory of planned behaviour* (TPB)—the most widely researched behavioural model—that actual behaviour is driven by the *intention* to perform an action. Intention, in turn, is predicted by three factors: first, *attitude*, which is formed by knowledge and beliefs about the behaviour in question and its outcome effects; second, *subjective norms*, shaped through a person's perception of how significant others think about the behaviour; and third, *perceived behaviour control* (PBC), which is derived from beliefs about the ease or difficulty of performing the behaviour. In the TPB, risk perception can be seen as one determinant of peoples' attitude towards the behaviour, but as with other models such as the HAPA, the theory does not imply that risk perception alone can form a person's intention to act nor explain actual health-related behaviour on its own.

In summary, the notion that risk perception influences health-related behaviour is prevalent in both theoretical behavioural science and health communication practices. Still, in health behaviour theories risk perception is generally seen as only one out of several relevant behavioural determinants, and its individual predictive quality for explaining behaviour has been challenged (see Brewer et al., 2007; Leventhal, Kelly, & Leventhal, 1999). Therefore, the goal of this chapter is to shed some light on the question of whether the construct of risk perception on its own is indeed a relevant predictor of health-related behaviour according to scientific evidence. To this end, we first outline the different dimensions of risk perception and discuss some methodological issues, accompanied by findings related to the effect of risk perception on health risks. Afterwards, we review the literature on the relationship between risk perception and safety-enhancing health behaviours.

## 7.2 Dimensions of Risk Perception

To understand how people's risk perception influences their behaviour, it is necessary to consider the respective dimensions of risk perception, as it is a rather broad and unspecific term. Researchers have adopted different categorisation systems. In the earlier literature and nearly all the existing models of health behaviour, risk perception is conceived as a construct with only two components: *perceived likelihood* and *perceived severity* (Brewer et al., 2007; van der Pligt, 1996). Perceived likelihood is the subjectively determined probability of experiencing a negative outcome in relation to one's behaviour (e.g., "Given that you did not clean your hands before patient contact, what would you say is the likelihood that dangerous pathogens are transmitted via your hands onto the patients?"). Perceived severity represents the assumed degree of harm arising from the negative outcome of one's behaviour (e.g., "How severe would the consequences for a patient be if he or she got infected with dangerous pathogens transmitted via unclean hands?").

In their meta-analysis, Brewer et al. (2007) argue for the additional distinction between *perceived likelihood of harm* and *perceived susceptibility to harm*. The latter is often used synonymously with perceived likelihood but actually is a person's constitutional vulnerability to harm (e.g., "I am more vulnerable than other people like me to being infected with a dangerous pathogen"). According to the authors, these dimensions are overlapping, as perceived susceptibility to a health hazard should influence the likelihood of being harmed by the hazard, but still logically distinct, as being vulnerable to harm does not by itself mean that the absolute probability of that harm is large (Brewer et al., 2007).

Earlier research on human risk behavior—which includes health-related actions—assumed that risky decision-making is a purely cognitive process (Loewenstein, Weber, Hsee, & Welch, 2001). In expected-utility theories, it is implied that an individual's cognitive assessment of the likelihood and severity of potential negative outcomes predicts risky choices (e.g., getting treatment for an illness or not) while emotions are not seen as an integral part of human decision-making. However, subsequent research has shown that affective processes indeed play a fundamental part in risk perception, decisions, and behavior (Loewenstein et al., 2001; Slovic, 2010; Slovic & Peters, 2006; see also Raue & Schneider, Chap. 5). Most risk perceptions in daily life do not arise out of a deliberate cognitive procedure but rather through automatic and intuitive processes. One characteristic of these automatic and intuitive processes is that they are strongly influenced by feelings, which is an extremely fast and resource-efficient way to make judgements and decisions in complex and potentially dangerous situations (Slovic & Peters, 2006). People often rely on these feelings and use them as a source of information to guide their decisions (formally known as the *affect heuristic,* see Slovic, Finucane, Peters, & MacGregor, 2002). Focusing on the mechanisms of affective reactions on peoples' risk perception is especially important in the health and safety domain, considering the range of feelings which are accompanied by many health-related events such as the anxiety of receiving a positive result after a cancer screening (Peters, McCaul, Stefanek, & Nelson, 2006). People might utilize their feelings (e.g., a healthcare workers' fear of transmitting pathogens onto a patient and causing an infection) as information to assess their own risk of being affected by a negative health outcome (e.g., a healthcare worker's perceived likelihood of transmitting dangerous pathogens onto a patient). In their meta-analysis, Sheeran et al. (2014) consequently proposed a more refined distinction between elements of human risk perception,[1] namely *risk perception as likelihood* (they do not differentiate between perceived likelihood and perceived susceptibility), *perceived severity, anticipatory emotions*, and *anticipated*

[1]The authors used the term risk appraisal for their categorisation system, because they reserved the term risk perception only for the cognitive evaluation of a potential threat (i.e., likelihood). There seem to be a lack of taxonomical consensus in the research area investigating the relationship between risk perception and health-related behaviour. For easier comprehensibility, the term risk perception was adopted for the generalised concept throughout the present chapter.

*emotions*. Anticipatory emotions such as fear and anxiety, on the one hand, are immediate visceral responses to potential risks and therefore are negative affective components of risk assessment. On the other hand, anticipated emotions such as regret and guilt are usually not felt immediately in the presence of potential threats to a person's safety but are components of the expected consequences of a decision or behavior (e.g., "I would feel guilty if a patient acquired an infection as a result of me not cleaning my hands").

To this day, most studies that investigated the relationship between risk perception and health-related behaviors included only a subset of the above-mentioned dimensions of risk perception (Sheeran et al., 2014). Often these subsets included only cognitive aspects of risk perception (e.g., perceived likelihood and perceived severity) or only affective aspects of risk perception such as anticipated emotions. Studies have shown that the cognitive and affective components of a person's risk perception are related. However, to the best of our knowledge, empirical research shows that they only correlate moderately with each other (McCaul & Mullens, 2003), and both have the potential to explain variance in behavior above and beyond each other (Sheeran et al., 2014). As already mentioned, including affective besides cognitive aspects of risk perception seems to be especially important for research in the health and safety domain, as health-related behavioral decisions are often made in emotionally charged situations. Therefore, looking only at a subset of the elements of risk perception can lead to an underestimation of the predictive power of the concept being examined. This can be seen as the first methodological concerns in the research field. Further topics are discussed below.

## 7.3 Methodological Issues with Measuring Risk Perception

As mentioned above, one problem of interpreting and comparing research that investigates the relationship between risk perception and health-related behavior is that many studies conceptualize and test risk perception in different ways. Besides this, there are several other methodological concerns regarding the measurement of risk perception (see also Lermer et al., Chap. 6). We want to discuss several of the most common problems: unconditioned questions, not controlling for past behavior, relying exclusively on correlational data, and refraining from defining the perception-to-behavior hypothesis tested in the study.

One of the most severe issues when testing the influence of risk perception on health-related behaviors is to leave risk questions unconditioned on the behavior in question (Brewer et al., 2004, 2007; Weinstein & Nicolich, 1993). We will explain this issue with an example: In a study that measures risk perception of healthcare workers regarding transmitting pathogens via their hands, the following question might be asked: "What is the likelihood that dangerous pathogens are transmitted via your hands onto the patients?" One respondent might judge the probability of this risk low because she thinks that her actions do not play a role in spreading germs onto patients. Another person might rate the risk low because she intends to clean her hands before touching a patient but does recognize a potential transmission risk that is eliminated by her safety-enhancing behavior. The second judgement already includes the health-related behavior while the first does not. Therefore, even though both risk perception measures are low, this does not mean that both persons indeed perceive the transmission risk to be low. To avoid this bias, the risk-related question should always be conditioned on not acting (e.g., "Given that you did not clean your hands before patient contact, what would you say is the likelihood that dangerous pathogens are transmitted via your hands onto the patients?").

Van der Pligt (1996) lamented that many studies that measure the effect of risk perception on health-related behaviors do not control for a person's past behavior. The predictive power of past behavior for future actions is well-documented (e.g., Ouellette & Wood, 1998) and yet past behavior is not a standard component of major health-behavior theories. One study by Otten and Van der Pligt (1992) showed that risky past behavior (e.g., having sexual intercourse with multiple partners in the past two years) resulted in a heightened risk perception in regard to the perceived likelihood of being affected by a related hazard (such as getting infected with a sexually transmitted disease), and that people's risk perception was related to their expected risky future behavior. However, when controlling for both risk perception and risky past behavior, only risky past behavior remained a statistically significant predictor for expected risky future behavior. Still, many studies and meta-analyses (e.g., Brewer et al., 2007; McEachan, Conner, Taylor, & Lawton, 2011) that rely on correlations, which measure the relationship between a target metric such as risk perception and an observable output such as health- and safety-related behavior, do not control for past behavior (Sheeran et al., 2017). To avoid an overestimation of the effect of risk perception on health- and safety-related behavior, past behavior should be included in statistical models.

Another related issue in previous research regarding the association between risk perception and health-related behavior lies in the fact that many studies only assess correlative data (see Sheeran et al., 2014). Correlations between two variables can describe a statistical connection, but cannot provide information about the meaningfulness and direction of the effect (Field, Miles, & Field, 2012). In our established example, high perceived risk of transmitting dangerous pathogens could influence a person to clean her hands before touching a patient (i.e., "I believe that there is a high risk for transmitting pathogens via my hands; therefore, I need to clean my hands to reduce that risk"). However, it could also be that regularly cleaning one's hands could result in a heightened level of risk perception which can henceforth be used to rationalize the behavior (i.e., "I often clean my hands as I have been told to do, thus, I must indeed believe that there is a high risk for transmitting pathogens via my hands, otherwise I would not do it"). At best, correlative cross-sectional data—such as questionnaire data that only captures one point in time—can only tell if the person's perceived risk perception accurately reflects that individual's risk-taking behavior (Weinstein & Nicolich, 1993). But also, with correlative longitudinal studies (e.g., questionnaire data measured at two or more time points) investigating the link between risk perception and health- and safety-related behavior the possibility that one or more third variables such as past behavior are causing the effect cannot be ruled out (Sheeran et al., 2014). Studies that present correlative data should not be interpreted as explaining cause and effect. Therefore, they cannot answer the question if a higher level of risk perception leads to a reduction in risk-taking or an increase in safety-enhancing behavior. This should by no means underestimate the value of correlative research. Correlations can help to identify potentially relevant determinants of behavior, especially when studying new activities like people's uptake of the recently developed pre-exposure prophylaxis (PrEP) against HIV infections or to this day largely neglected behaviors such as hand hygiene when visiting a hospital. Still, the only studies that can help to explain causal direction are those in which risk perception has been successfully experimentally manipulated and the manipulation's impact on health-related behaviors subsequently measured.

When scholars intend to examine the association between constructs like risk perception and health- and safety-related behaviors (which is proposed by behavioral theories such as the health action process approach), different assumptions about this association can be tested, namely, the *accuracy hypothesis*, the *risk reappraisal hypothesis*, and the *behavior motivation hypothesis* (Brewer, Weinstein, Cuite, & Herrington, 2004). First, the accuracy hypothesis tests whether a person's risk perception at any given moment reflects that person's risk-taking behavior at the same time (Weinstein & Nicolich, 1993). The basic idea is, that "people who engage in risky behaviors have higher actual risk and should have higher perceived risk" and vice versa (Brewer et al., 2004, p. 126). For example, a researcher might be interested to know whether healthcare workers who believe they have a low or high risk of transmitting dangerous pathogens via their hands are truly correct in their assumption when considering their current hand hygiene behavior. The accuracy hypothesis describes the correlation between risk perception and risk-taking behavior at a given time and can be assessed through cross-sectional questionnaire data (i.e., data that is collected from many people at one time to test for between-subject differences, disregarding any changes over time). It cannot provide information about the causal direction between the two factors. However, according to Brewer et al. (2004), these kinds of results, which are commonly reported in studies, are often misinterpreted in the way that risk perception determines people's behavior.

The risk reappraisal assumption "describes the effects of changes in behavior on changes in perceived risk" (Brewer et al., 2004, p. 127). Essentially, it is a measure of how effective people believe their behavior (i.e., taking safety-enhancing actions such as hand hygiene or avoiding potentially harmful activities such as smoking) is in reducing their risk for harm. If the behavior in question is thought to decrease the risk of health hazards, then people who engage in that particular behavior should have a reduced risk perception afterwards. When linking it back to our hand hygiene example, if a physician thinks that hand hygiene is an effective method to prevent the spread of dangerous pathogens, she should consider herself at a lower transmission risk after adopting the habit of always cleaning her hands before patient contact. To test the reappraisal hypothesis, measures of risk perception from at least two points in time (before and after the adoption of the behavior) are needed.

The behavior motivation hypothesis tests what many scholars and practitioners are actually interested in—the assumption that people's risk perception has a causal effect on their risk-taking/safety-enhancing behavior (Brewer et al., 2004). In our example, this would mean the following: If an intervention is successful in increasing the risk perception of healthcare workers, that they are a potential vector for transmitting dangerous pathogens, this should lead to a higher rate of hand hygiene in healthcare facilities. As with the reappraisal hypothesis, testing the behavior motivation hypothesis needs at least a longitudinal and at best an experimental study design. Researchers need to know exactly what they want to measure and have to adopt an appropriate method to study this effect to avoid misinterpretations (see also Lermer, Streicher, & Raue, 2018).

In addition to the already mentioned methodological concerns, several other factors can influence the results of risk-perception-to-behavior studies. For instance, the answer format used in a questionnaire can affect the relationship between risk perception and health behavior. To this day, there is no consensus on how to assess risk perception, especially regarding the answer format (Lermer, Streicher, Sachs, & Frey, 2013), even though it has been long known that the scale on which risk questions are to be answered has an impact on people's risk assessment (for comprehensive reviews on this topic see Lermer et al., 2013, 2016, 2018 and Lermer et al., Chap. 6). In sum, the above-mentioned methodological issues should be kept in mind when interpreting the results of studies looking at the link between risk perception and health behavior. If scholars are unaware or inconsiderate of the problems caused by lower-quality research, a misguided conclusion might be drawn.

## 7.4 Research Findings on How Risk Perception Influences People's Health Behavior

Previously, we described the different dimensions of risk perception and some of the methodological issues which have been linked to risk research. Finally, we review the literature on the link between risk perception and health-related behaviors. The research field is vast, and therefore we will provide some general meta-analytic findings first before focusing on two safety-enhancing health behaviors: hand hygiene in hospitals and vaccination. Both have been well researched and have different behavioral dimensions attached to them.

While most of the research looked at a specific health-related activity, some meta-analyses incorporated a variety of behaviors to test the relationship between risk perception and risk-taking/safety-enhancing behavior. For instance, Floyd et al. (2000) considered 65 studies investigating the connection between two components of risk perception (perceived severity of and perceived susceptibility to a threat) and health-related behaviors belonging to one of three categories: (1) *ceasing* harmful actions like smoking, (2) *maintaining* beneficial activities such as continuing to exercise, and (3) *initiating* protective behaviors like wearing sunscreen. The authors accepted both intention and actual behavior as dependent measures, but also conducted separate analyses for both variables. The reason for the latter was that intention does not always translate into actual behavior, a phenomenon formally known as the 'intention-behavior-gap' (Sniehotta, Scholz, & Schwarzer, 2005). Overall, they found that the effect size[2] between risk perception (perceived susceptibility to and perceived severity of a threat combined into one predictor variable) and health behavior (intention and actual behavior combined) was moderate ($d_+ = 0.54$). The link between risk perception and the health-related behavioral variables was stronger for the intention to engage in an activity ($d_+ = 0.56$) compared to actual behavior ($d_+ = 0.41$). The authors also found comparable small-to-medium-sized effects between the two dimensions of risk appraisal when looking at the different behavioral categories (i.e., ceasing harmful actions and initiating protective behavior; due to a lack of studies, they could not calculate the effect sizes for maintaining beneficial activities). However, the authors included predominantly correlative studies in their meta-analysis, which prohibits a cause-and-effect interpretation of the results. Milne et al. (2000) conducted a very similar meta-analysis on the relationship between three risk perception components on several health risk detection (e.g., mammography) and prevention (e.g., sunscreen usage) behavioral intentions and actual concurrent and subsequent behaviors. The dependent variable, concurrent behavior, was gathered through cross-sectional studies (i.e., data measured at only one time), while the measuring of subsequent behavior was conducted through longitudinal (i.e., data measured at two or more time points) and experimental data. Besides perceived susceptibility and perceived severity, the anticipatory emotion of *fear* was added to the risk perception concept. The authors reported a small association between severity and behavioral intention ($r_+ = 0.10$) and small to moderate correlations between susceptibility and intention ($r_+ = 0.16$) as well as fear and intention ($r_+ = 0.20$). They also found small-to-medium-sized relationships between risk perception and concurrent behaviors (fear: $r_+ = 0.26$; susceptibility: $r_+ = 0.13$; and severity: $r_+ = 0.10$) and only perceived susceptibility turned out to show a significant but small correlation with subsequent behaviors ($r_+ = 0.12$). Overall, the researchers found that

---

[2]An effect size is an objective and standardised measures of the magnitude of the observed statistical effect. The main benefits of having a standardised measure are, that effect sizes of different studies can be compared at the same level and that it allows an inference about the meaningfulness/importance of the effect. There are several standard measures of effect sizes. The most commonly used are Cohen's d which indicates mean differences between two groups and the Pearson correlation coefficient r, which is an expression of the strength of the relationship between variables. Cohen (1988) proposed indications on how to interpret the effect sizes. For the Cohen's d: d = 0.2 (small effect), d = 0.5 (medium effect), and d = 0.8 (large effect). For the Pearson correlation coefficient r: r = .1 (small effect), r = .3 (medium effect), and r = .5 (large effect).

risk perception components (especially fear) were a significant determinant of health behaviors. However, while both correlative and experimental data were included in this systematic review, the majority of studies relied on a correlative design.

One subsequent meta-analysis of 208 studies focused only on experimental research that tested the link between risk perception and health-related behavior (Sheeran et al., 2014). The authors looked at four of the above-mentioned dimensions of risk perception (perceived likelihood, perceived severity, anticipatory emotions, and anticipated emotions) and their associations with a diverse set of behavioral intentions and actual health- and safety-related behaviors. Overall, interventions that heightened risk perception (combined variable from all studies that increased at least one dimension of risk perception) had a significant yet only small effect on intention ($d_+ = 0.31$) and behavior ($d_+ = 0.23$). The authors also tested if a heightened risk perception was more effective in changing behavior when there was an increase in more than one of the dimensions of risk perception within the same intervention. They found that the combined effect of the increased perceived likelihood of harm and heightened perceived severity of the negative consequences ($d_+ = 0.36$) was greater compared to studies in which perceived severity was not amplified ($d_+ = 0.16$). The same was true for studies that succeeded in increasing both perceived likelihood and anticipatory emotion ($d_+ = 0.22$ compared to $d_+ = 0.10$ when anticipatory emotion was not increased). However, perceived severity was not much more effective in changing people's behavior when anticipatory emotion was increased ($d_+ = 0.41$ when both were increased compared to $d_+ = 0.32$ when only severity was raised). Overall, the results from Sheeran et al. support the notion that the different dimensions of risk perception have distinct features in influencing health-related behavior outcomes. Interventions that manage to increase several elements of risk perception are more effective in changing people's behavior. Still, the authors concluded that the direct effects of risk perception (single dimensions as well as combined variables) on health-related behavior were rather small. One factor which helps to explain this is that effects seem to depend on the type of activity in question. Increases in risk perception had the most substantial positive effect on sun protection ($d_+ = 0.40$), vaccination ($d_+= 0.33$), and diet ($d_+ = 0.34$). However, risk perception had no statistically significant effect on other health-related activities like obtaining diagnostic medical tests, dental hygiene, or alcohol consumption.

With this in mind, we took a closer look at the association between risk perception and two distinct health behaviors: hand hygiene in healthcare facilities and vaccination. We chose these two behaviors because they differ sharply regarding the frequency of occurrence and the ease of execution. Hand hygiene within or outside healthcare facilities is a routine behavior that people generally perform many times every day, while getting vaccinated occurs rather infrequently, about once a year. Hand hygiene is also very quick, easy, and painless to execute – especially in hospitals where alcohol-based hand-rub dispensers are ubiquitous – while people usually have to arrange an appointment with their physician to get vaccinated, which can be uncomfortable by itself.

### 7.4.1 Hand Hygiene Behavior

Adequate hand hygiene is one of the most important and effective methods to prevent healthcare-associated infections, which pose a risk to the health of millions of patients around the globe (WHO, 2009). Still, compliance with hand hygiene guidelines in hospitals is suboptimal in most healthcare facilities worldwide (Erasmus, Daha, Brug, Richardus, Behrendt, Vos, et al., 2010). Several factors that can influence people's hand hygiene behavior have been identified and well-researched, including social norms (Gaube, Tsivrikos, Dollinger, & Lermer, 2018b), feedback, memory, and attention (Fuller, Besser, Savage, McAteer, Stone, & Michie, 2014; Gaube, Lermer, & Fischer, 2018a) among many others. Somewhat surprisingly, risk perception has received comparatively little attention. In a study among physicians, Pittet et al. (2004) found that most physicians (85%) were aware of the risk of transmitting pathogens to patients if they did not clean their hands, yet on average the observed compliance with guidelines was only 57%. When only looking at the variables of risk perception and hand hygiene compliance, the level of risk perception ("Does non-compliance with hand hygiene imply a risk of cross-transmission to the patient?") was a significant predictor for adherence to guidelines. However, when the authors introduced a more complex statistical model that controlled for other factors like attitude and norms, physicians' risk perception was no longer significantly associated with adherence to guidelines. Other studies also did not find a significant correlation between healthcare workers' risk perception and hand hygiene behavior or intention to act (e.g., Pessoa-Silva, Posfay-Barbe, Pfister, Touveneau, Perneger, & Pittet, 2005; Sax, Uckay, Richet, Allegranzi, & Pittet, 2007). One possible explanation for why risk perception has not been found to have a significant influence on hand hygiene compliance might lie in the nature of the behavior. Healthcare workers have to clean their hands numerous times a day. Therefore, for regular practitioners, hand hygiene might become a highly automatized behavior that does not involve deliberate thinking (Diefenbacher, Sassenrath, Siegel, Grünewald, & Keller, 2012). All the studies mentioned above only looked at the cognitive aspects of risk perception (i.e., likelihood and severity). However, it might be of more interest to test in what ways the more intuitive affective components of risk perception might influence hand hygiene behavior. To our knowledge, this has not been studied.

Still, the cognitive aspects of risk perception could be helpful to explain variations in hand hygiene behavior in hospitals among groups of people for whom this is not a routine activity. For instance, hospital visitors might not be accustomed to using alcohol-based hand-rub dispensers for their personal hand hygiene and so must deliberately consider whether and where they should use the dispenser in the healthcare facility. In a recently conducted survey among hospital visitors based on the health action process approach, perceived likelihood of harm for patients and oneself when hand hygiene was omitted was the only significant correlate with observed hand hygiene behavior in the hospitals' entrance hall (Gaube et al., 2018a).

From a methodological standpoint, one positive fact is that almost all risk perception measures of the above-mentioned studies on hand hygiene have been conditioned on the behavior (see under '*Methodological issues with measuring risk perception*'). This means that the risk perception questions asked about the likelihood of spreading pathogens or severity of harm when hand hygiene was omitted. However, only our study (Gaube et al., 2018a) seemed to have controlled for past behavior. It should also be noted that all of the studies that examined the link between risk perception and hand hygiene behavior relied on cross-sectional correlative data that cannot answer whether higher levels of risk perception trigger hand hygiene behavior or vice versa. More research is needed to investigate in what ways the different elements of risk perception influence hand hygiene behavior in healthcare settings. Additional knowledge about the factors influencing hand hygiene could help in advancing interventions to improve patient safety in healthcare facilities.

## 7.5 Vaccination

According to the World Health Organization, immunization through vaccination is one of the most successful and cost-effective health interventions available. Through vaccinations, countless lives around the world are saved, and with its help some formerly devastating diseases such as smallpox have been eradicated (WHO, 2013). Still, the uptake of vaccinations against various diseases is below the level's experts pursue (Dubé, Laberge, Guay, Bramadat, Roy, & Bettinger, 2013). Risk perception has been proposed to be one of the key predictors for people's vaccination behavior, and it has been studied in greater depth than hand hygiene compliance. In their meta-analysis, Brewer et al. (2007) suggested that risk perception might be more relevant for behaviors that aim to reduce or eliminate a specific health threat such as vaccination and less relevant for general health promotion behaviors like regular exercising. Among the 34 studies included in the analysis, the authors found that the perceived likelihood of being harmed by the to-be-vaccinated-against illness was a significant predictor for vaccination uptake ($r_+ = 0.26$). The same was found for perceived susceptibility to the disease ($r_+ = 0.24$) and severity of harm caused by the disease ($r_+ = 0.16$). These correlations were larger then previous research had suggested. Brewer et al. (2007) concluded that the results confirmed the importance of risk perception as an integral part of many health behavior theories. Additionally, they reported that prospective studies—which measured the level of risk perception before the respondent got vaccinated or not—and studies using high-quality measures of risk perception (e.g., risk questions conditioned on not having been vaccinated), yielded stronger effects in comparison to cross-sectional research. This again makes a strong statement for the importance of methodologically rigorous health behavior research.

However, the meta-analysis did not include affective components of risk perception. In a prospective study with two time-points (before and after the availability of the seasonal influenza vaccine) several components of risk perception were compared on their predictive quality for getting a flu shot (Weinstein, Kwitel,

McCaul, Magnan, Gerrard, & Gibbons, 2007). The researchers found that anticipated regret ($r = 0.45$) was the best single predictor for vaccination behavior, followed by the feeling to be at risk ($r = 0.44$) to the flu. Overall, the prognostic quality of affect-driven components of risk perception was superior to the cognitive probability judgements of risk (risk magnitude judgement $r = 0.30$; beliefs about risk $r = 0.25$). These results confirm the significance of including affective components as an essential part of general risk perception when looking at the link between risk perception and health- and safety-related activities.

Two successional systematic reviews on immunization behavior have generally validated previous findings. Bish et al. (2011) found that both an individual's feeling of being at risk of pandemic influenza and the feeling of worry about pandemic influenza were positively linked to intention and uptake of influenza vaccination. Besides risk perception, it was reported that immunization behavior was also influenced by people's perception of the vaccine's effectiveness in preventing the target disease and its perceived safety. Additionally, the authors claimed that social norms and past vaccination behavior were good predictors of intention and future immunization behavior. A meta-analysis investigating the factors that affect the acceptance of human papillomavirus (HPV) vaccination amongst men found anticipatory regret of not being vaccinated ($r_+ = 0.27$) and perceived susceptibility to HPV ($r_+ = 0.25$) to be among the most potent correlates (Newman, Logie, Doukas, & Asakura, 2013). However, it was not the perceived risk of the HPV infection but the perceived benefits of getting an HPV vaccination ($r_+ = 0.51$) that was the strongest single correlate of acceptability. The reader should note that in all the studies mentioned above, perceived severity of the health risk was a significant predictor of immunization behavior but at a lower magnitude. Also, many studies included in the systematic reviews relied solely on cross-sectional correlational data. Therefore, one should be cautious in interpreting the results in respect to cause and effect. In summation, research indicates that the various aspects of risk perception are linked to vaccination behavior. Affective components of risk perception seem to be especially relevant predictors of vaccination uptake. However, people's risk perception is far from being explanatory for all variance in immunization practices.

## 7.6 Summary and Conclusion

Many governmental and organizational interventions designed to change health-related behaviors try to intensify people's risk perception of adverse outcomes. Targeting risk perception should help people to vacate harmful activities such as smoking or uptake/improve preventive, safety-enhancing behaviors such as adequate hand cleaning in hospitals or getting vaccinated. Risk perception is also explicitly or implicitly an integral part of many health behavior theories that try to explain and predict human behavior. The goal of the present book chapter was to provide a brief overview of the literature that looks at the link between people's risk perception and their health- and safety-related behavior to see if the prevalence of the concept in both theory and practice is justifiable through empirical results. First, research has shown that risk perception is not a one-dimensional construct but instead consists of several cognitive and affective elements. On the cognitive side, the perceived likelihood of being harmed and the perceived severity of the potential health hazard predominate the literature. On the affective side, there is a distinction between *anticipatory* emotions and *anticipated* emotions in regard to the risk. Second, several methodological issues have been discussed that can affect how study results should be interpreted. If risk perception is measured with questions that are unconditioned to the behavior in question, the respondents might assess their risk from entirely different standpoints (e.g., low risk perception of adverse consequences because protective behavior has been performed or is planned versus perception of being at low risk and believing that no protective behavior is needed) which can bias the study results. Also, the link between risk perception and health-related activities can be overestimated if the researchers do not control for past behavior. Result misinterpretation can also occur when correlative cross-sectional study design is used to test research questions that in reality can only be answered through an experimental or at least longitudinal research design – such as determining a causal relationship between a person's risk perception and health-related behavior. It is crucial that scholars pre-define the perception-to-behavior hypothesis they intend to test and adopt an appropriate research design afterwards. Otherwise, it is not clear how strong the link between risk perception and behavior is or if the relation is moderated by a third variable.

Finally, the goal of the present chapter was to examine whether the construct of risk perception on its own is indeed a relevant predictor of health-related behavior according to scientific findings. Overall, the literature suggests that people's risk perception of a specific health threat indeed influences their behavior (normally in a risk-reducing direction). Therefore, it seems to be empirically justifiable that risk perception holds a prominent position in both health behavior theory and health behavior change interventions. However, researchers and practitioners should be aware that the magnitudes of the correlations are generally of medium size. Still, even moderate effects can be of substantial population-wide importance. For instance, Brewer et al. (2007) estimated that a correlation of .28 between risk perception and vaccination behavior would correspond to a vaccination rate of 36% among people with low risk perception, compared to a vaccinate rate of 64% among people with high risk perception. This example also indicates that even interventions that manage to increase people's risk perception towards a health hazard only slightly might have a profound applied effect on the target population's health behavior.

When looking closer at the different dimensions of risk perception, the majority of studies showed that the affective components of risk perception—especially anticipated regret—had a stronger predictive power compared to more cognitive aspects such as perceived likelihood to harm and severity of a health hazard. These findings indicate that commonplace information campaigns intended to increase people's risk perception of health hazards in terms of likelihood, severity, and susceptibility might not be the most efficient tools for health behavior modification if they do not change people's feelings of risk simultaneously. Therefore, health behavior change interventions should ideally target both the affective and cognitive elements of people's risk perception to generate optimal results.

Campaign designers should carefully consider how to accomplish this goal and ideally test audience's reactions to the material before rolling it out to the wider public. One good example for pre-testing campaign messages was published by Cox et al. (2014). In an experimental setting, they tested the effect of different materials on mothers' intention to get their daughters vaccinated against HPV. All participating mothers in the control and the experimental group received some general information about HPV infection and immunization. Next, participants in the experimental group were exposed to additional material that targeted cognitive (likelihood) and affective (anticipated regret) risk perception components. Mothers in the experimental condition were presented with information about the statistical likelihood of girls being affected by cervical cancer either in a text-only or graphically illustrated fashion. Afterwards, about half of these women were asked about their anticipated regret if their daughter developed cervical cancer and hadn't been vaccinated while the other half was not asked this question. Results indicated that triggering anticipated regret indeed led to the highest levels of vaccination intention, but only in the group that received graphically illustrated statistical information. The authors concluded that targeting anticipated regret is only an effective health-related behavior change tool if the intervention's recipient was able to easily comprehend (e.g., through graphically illustrated statistical messages) potential risks of the health threat. These findings do not only support previous research findings suggesting that targeting affective risk dimensions are very effective but also highlights the complexity of designing effective health behavior change interventions.

Overall, most scholars agree that triggering risk-related emotions can increase the effectiveness of health and safety campaigns and that the usage of pictures and other visual materials is an effective way to deliver a persuasive message and evoke emotions in the audience (Joffe, 2008). However, campaign designers should consider the facilitation of affective risk perception components carefully, as it has the potential to backfire in terms of prompting a defense response, which means that individuals evade or ignore the message instead of changing their behavior to avoid the risk of negative health consequences. Defense responses are more likely when strong negative emotions such as fear are triggered (Leshner, Bolls, & Thomas, 2009). Therefore, the prevalence of fear-evoking headlines such as the one mentioned at the chapter's beginning ("*Killer germs: Death from the hospital*"; von Leszczynski, 2016) might not only run counter to people's striving for safety and security, but also be of little assistance in changing people's health- and safety-related behaviors.

## References

Ajzen, I. (1991). The theory of planned behavior. *Organizational Behavior and Human Decision Processes, 50*(2), 179–211.

Bish, A., Yardley, L., Nicoll, A., & Michie, S. (2011). Factors associated with uptake of vaccination against pandemic influenza: A systematic review. *Vaccine, 29*(38), 6472–6484.

Brewer, N. T., Chapman, G. B., Gibbons, F. X., Gerrard, M., McCaul, K. D., & Weinstein, N. D. (2007). Meta-analysis of the relationship between risk perception and health behavior: The example of vaccination. *Health Psychology, 26*(2), 136–145.

Brewer, N. T., Weinstein, N. D., Cuite, C. L., & Herrington, J. E. (2004). Risk perceptions and their relation to risk behavior. *Annals of Behavioral Medicine, 27*(2), 125–130.

Cohen, J. (1988). *Statistical power analysis for the behavioral sciences* (2nd ed.). New York: Academic Press.

Cox, D., Sturm, L., & Cox, A. D. (2014). Effectiveness of asking anticipated regret in increasing HPV vaccination intention in mothers. *Health Psychology, 33*(9), 1074–1083.

Diefenbacher, S., Sassenrath, C., Siegel, A., Grünewald, M., & Keller, J. (2012). Implizite Einstellung zur Händehygiene als relevanter Prädiktor von Händehygieneverhalten. *Hygiene & Medizin, 38*(11), 448–455.

Dubé, E., Laberge, C., Guay, M., Bramadat, P., Roy, R., & Bettinger, J. A. (2013). Vaccine hesitancy: An overview. *Human Vaccines & Immunotherapeutics, 9*(8), 1763–1773.

Erasmus, V., Daha, T. J., Brug, H., Richardus, J. H., Behrendt, M. D., Vos, M. C., et al. (2010). Systematic review of studies on compliance with hand hygiene guidelines in hospital care. *Infection Control and Hospital Epidemiology, 31*(3), 283–294.

Field, A., Miles, J., & Field, Z. (2012). *Discovering Statistics Using R*. London: SAGE Publications Ltd.

Floyd, D. L., Prentice-Dunn, S., & Rogers, R. W. (2000). A meta-analysis of research on protection motivation theory. *Journal of Applied Social Psychology, 30*(2), 407–429.

Fuller, C., Besser, S., Savage, J., McAteer, J., Stone, S., & Michie, S. (2014). Application of a theoretical framework for behavior change to hospital workers' real-time explanations for noncompliance with hand hygiene guidelines. *American Journal of Infection Control, 42*(2), 106–110.

Gaube, S., Lermer, E., & Fischer, P. (2018a). *Identifying psychological factors that influence hospital visitor's hand hygiene behavior*. Manuscript in preparation.

Gaube, S., Tsivrikos, D., Dollinger, D., & Lermer, E. (2018b). How a smiley protects health: A pilot intervention to improve hand hygiene in hospitals by activating injunctive norms through emoticons. *PLoS ONE, 13*(5), e0197465.

Hastings, G., Stead, M., & Webb, J. (2004). Fear appeals in social marketing: Strategic and ethical reasons for concern. *Psychology & Marketing, 21*(11), 961–986.

Joffe, H. (2008). The power of visual material: Persuasion, emotion and identification. *Diogenes, 55*(1), 84–93.

Leshner, G., Bolls, P., & Thomas, E. (2009). Scare'em or disgust'em: The effects of graphic health promotion messages. *Health Communication, 24*(5), 447–458.

Lermer, E., Raue, M., & Frey, D. (2016). Risikowahrnehmung und Risikoverhalten. In D. Frey & H.-W. Bierhoff (Eds.), *Enzyklopädie der Psychologie—Soziale Motive und soziale Einstellungen—Sozialpsychologie*. Göttingen: Hogrefe.

Lermer, E., Streicher, B., & Raue, M. (2018). Measuring subjective risk estimates. In M. Raue, E. Lermer, & B. Streicher (Eds.), *Psychological Perspectives on Risk and Risk Analysis: Theory, Models and Applications*. New York: Springer.

Lermer, E., Streicher, B., Sachs, R., & Frey, D. (2013). How risky? The impact of target person and answer format on risk assessment. *Journal of Risk Research, 16*(7), 903–919.

Leventhal, H., Kelly, K., & Leventhal, E. A. (1999). Population risk, actual risk, perceived risk, and cancer control: A discussion. *Journal of the National Cancer Institute Monographs, 25*, 81–85.

Loewenstein, G. F., Weber, E. U., Hsee, C. K., & Welch, N. (2001). Risk as feelings. *Psychological Bulletin, 127*(2), 267–286.
McCaul, K. D., & Mullens, A. B. (2003). Affect, thought and self-protective health behavior: the case of worry and cancer screening. In J. Suls & K. A. Wallston (Eds.), *Social psychological foundations of health and illness*. Malden (MA): Blackwell Publishing Ltd.
McEachan, R. R., Conner, M., Taylor, N. J., & Lawton, R. J. (2011). Prospective prediction of health-related behaviours with the theory of planned behavior: A meta-analysis. *Health Psychology Review, 5*(2), 97–144.
Milne, S., Sheeran, P., & Orbell, S. (2000). Prediction and intervention in health-related behavior: A meta-analytic review of protection motivation theory. *Journal of Applied Social Psychology, 30*(1), 106–143.
Newman, P. A., Logie, C. H., Doukas, N., & Asakura, K. (2013). HPV vaccine acceptability among men: A systematic review and meta-analysis. *Sex Transm Infect, 89*(7), 568–574.
Otten, W., & Van der Pligt, J. (1992). Risk and behavior: The mediating role of risk appraisal. *Acta Psychologica, 80*(1–3), 325–346.
Ouellette, J. A., & Wood, W. (1998). Habit and intention in everyday life: The multiple processes by which past behavior predicts future behavior. *Psychological Bulletin, 124*(1), 54–74.
Pessoa-Silva, C. L., Posfay-Barbe, K., Pfister, R., Touveneau, S., Perneger, T. V., & Pittet, D. (2005). Attitudes and perceptions toward hand hygiene among healthcare workers caring for critically ill neonates. *Infection Control and Hospital Epidemiology, 26*(3), 305–311.
Peters, E., McCaul, K. D., Stefanek, M., & Nelson, W. (2006). A heuristics approach to understanding cancer risk perception: Contributions from judgment and decision-making research. *Annals of Behavioral Medicine, 31*(1), 45–52.
Pittet, D., Simon, A., Hugonnet, S., Pessoa-Silva, C. L., Sauvan, V., & Perneger, T. V. (2004). Hand hygiene among physicians: Performance, beliefs, and perceptions. *Annals of Internal Medicine, 141*(1), 1–8.
Reynolds, T. (2001). News headlines feed on fear of cancer risk, experts say. *Journal of the National Cancer Institute, 93*(1), 9–11.
Rogers, R. W. (1983). Cognitive and psychological processes in fear appeals and attitude change: A revised theory of protection motivation. In J. Cacioppo & R. Petty (Eds.), *Social psychophysiology*. New York: Guilford Press.
Rosenstock, I. (1966). Why people use health services. *The Milbank Memorial Fund Quarterly, 44,* 94–127.
Ruiter, R. A., Kessels, L. T., Peters, G. J. Y., & Kok, G. (2014). Sixty years of fear appeal research: Current state of the evidence. *International Journal of Psychology, 49*(2), 63–70.
Sax, H., Uckay, I., Richet, H., Allegranzi, B., & Pittet, D. (2007). Determinants of good adherence to hand hygiene among healthcare workers who have extensive exposure to hand hygiene campaigns. *Infection Control and Hospital Epidemiology, 28*(11), 1267–1274.
Schwarzer, R. (2008). Modeling health behavior change: How to predict and modify the adoption and maintenance of health behaviors. *Applied Psychology, 57*(1), 1–29.
Schwarzer, R., Lippke, S., & Luszczynska, A. (2011). Mechanisms of health behavior change in persons with chronic illness or disability: The health action process approach (HAPA). *Rehabilitation Psychology, 56*(3), 161–170.
Sheeran, P., Harris, P. R., & Epton, T. (2014). Does heightening risk appraisals change people's intentions and behavior? A meta-analysis of experimental studies. *Psychological Bulletin, 140* (2), 511–543.
Sheeran, P., Klein, W. M., & Rothman, A. J. (2017). Health behavior change: Moving from observation to intervention. *Annual Review of Psychology, 68,* 573–600.
Slovic, P. (2010). *The feeling of risk: New perspectives on risk perception*. Oxford, UK: Routledge.
Slovic, P., Finucane, M., Peters, E., & MacGregor, D. G. (2002). Rational actors or rational fools: Implications of the affect heuristic for behavioral economics. *The Journal of Socio-Economics, 31*(4), 329–342.

Slovic, P., & Peters, E. (2006). Risk perception and affect. *Current Directions in Psychological Science, 15*(6), 322–325.
Sniehotta, F. F., Scholz, U., & Schwarzer, R. (2005). Bridging the intention–behavior gap: Planning, self-efficacy, and action control in the adoption and maintenance of physical exercise. *Psychology & Health, 20*(2), 143–160.
van der Pligt, J. (1996). Risk perception and self-protective behavior. *European Psychologist, 1*(1), 34–43.
von Leszczynski, U. (2016). Killerkeime: Der Tod aus dem Krankenhaus. *Südwest Presse*. Retrieved from https://www.swp.de/panorama/der-tod-aus-dem-krankenhaus-23162265.html.
Weinstein, N. D., Kwitel, A., McCaul, K. D., Magnan, R. E., Gerrard, M., & Gibbons, F. X. (2007). Risk perceptions: Assessment and relationship to influenza vaccination. *Health Psychology, 26*(2), 146–151.
Weinstein, N. D., & Nicolich, M. (1993). Correct and incorrect interpretations of correlations between risk perceptions and risk behaviors. *Health Psychology, 12*(3), 235–245.
WHO (2009). WHO Guidelines on Hand Hygiene in Health Care: A Summary. First Global Patient Safety Challenge Clean Care is Safer Care. Retrieved from http://www.who.int/gpsc/5may/tools/9789241597906/en/.
WHO (2013). Global Vaccine Action Plan 2011–2020. Retrieved from http://www.who.int/immunization/global_vaccine_action_plan/GVAP_doc_2011_2020/en/.
Witte, K. (1992). Putting the fear back into fear appeals: The extended parallel process model. *Communications Monographs, 59*(4), 329–349.

# Chapter 8
# Perception of Aviation Safety

**Robert Mauro**

**Abstract** Aviation professionals and the general population have different perceptions of aviation safety. In the general population, there is a conflict between facts and fears. People know that commercial aviation is extremely safe, but they do not always feel safe and often make decisions based on those feelings. For aviation professionals, the conflict is between facts and facts, with managers and regulators struggling to weigh costs and benefits with risks in making safety related decisions. In this chapter the similarities and differences between lay and professional perceptions of aviation safety, the factors that affect these perceptions, and the effect of these perceptions on decisions are explored.

**Keywords** Perceived safety · Perceived risk · Aviation safety · Airline safety · Risk assessment

> Aviation in itself is not inherently dangerous. But to an even greater degree than the sea, it is terribly unforgiving of any carelessness, incapacity or neglect.
>
> —*Captain A. G. Lamplugh, British Aviation Insurance Group, London. c. early 1930's.*

## 8.1 Introduction

By most measures, commercial aviation is extremely safe. According to the International Air Transport Association (International Air Transport Association 2015), in 2014 more than 3.3 billion people traveled by commercial airliner on 38 million flights. In that year, there were 73 commercial aircraft accidents, 12 of them

R. Mauro (✉)
Decision Research, 1201 Oak St. Suite 200, Eugene, OR 97401, USA
e-mail: mauro@uoregon.edu

R. Mauro
Psychology Department, University of Oregon, 1227, Eugene, OR 97403, USA



M. Raue et al. (eds.), *Perceived Safety*, Risk Engineering,
https://doi.org/10.1007/978-3-030-11456-5_8

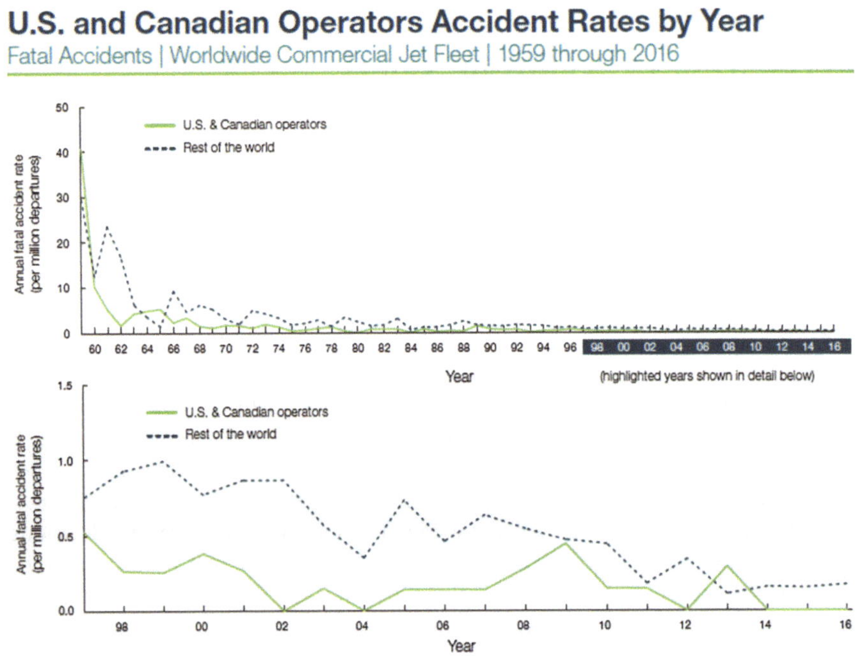


**Fig. 8.1** Fatal accidents by year

fatal, resulting in 641 deaths, which corresponds to 0.23 deaths from commercial aircraft accidents per million flights. Put another way, in 2014 the probability of boarding a flight anywhere in the word that resulted in a fatality was slightly less than 1 in 3,000,000. In the western world, commercial aviation is even safer. For example, in the United States (US), in 2014, 764 million passengers flew on commercial airliners. There were 13 commercial aviation accidents. As a result of these accidents there were no fatalities or serious injuries. As Fig. 8.1 demonstrates, this level of safety is not an aberration. No matter how one parses the data, commercial air travel is extraordinarily safe. You are very unlikely to die in a commercial aviation accident. According to the National Safety Council (2015), the lifetime odds of dying as a result of an aviation (or space) accident are 1:8,015 compared to 1:112 for motor vehicle accidents (see Fig. 8.2).[1]

For the most part, people appear to *know* that commercial aviation is extremely safe. For example, in a recent national poll of US residents, 73% reported that they thought that the overall safety record of commercial airline travel was "good" or "excellent" (CBS 2014). In another poll, 76% of a US national sample reported that they believed that air travel in the United States was "somewhat" or "very" safe

[1]How these statistics are calculated and determining what the most appropriate comparisons are is a task for another paper. For example, should one compare fatalities per person, trip taken, hours in transit, or miles travelled? Each statistic is appropriate for a different purpose.

(Ipsos-McClatchly 2010). However, this does not prevent people from *feeling* anxious about flying. For example, in a recent poll of US residents, only 41% of the respondents reported that they were "not at all afraid" to fly (Frankovic 2015). In another recent national poll of US residents, only 40% reported that they were "not afraid at all" of flying while 35% said that they were "somewhat afraid" or "very afraid" (Gallup 2006). Understanding the interplay between *cognitions* about safety

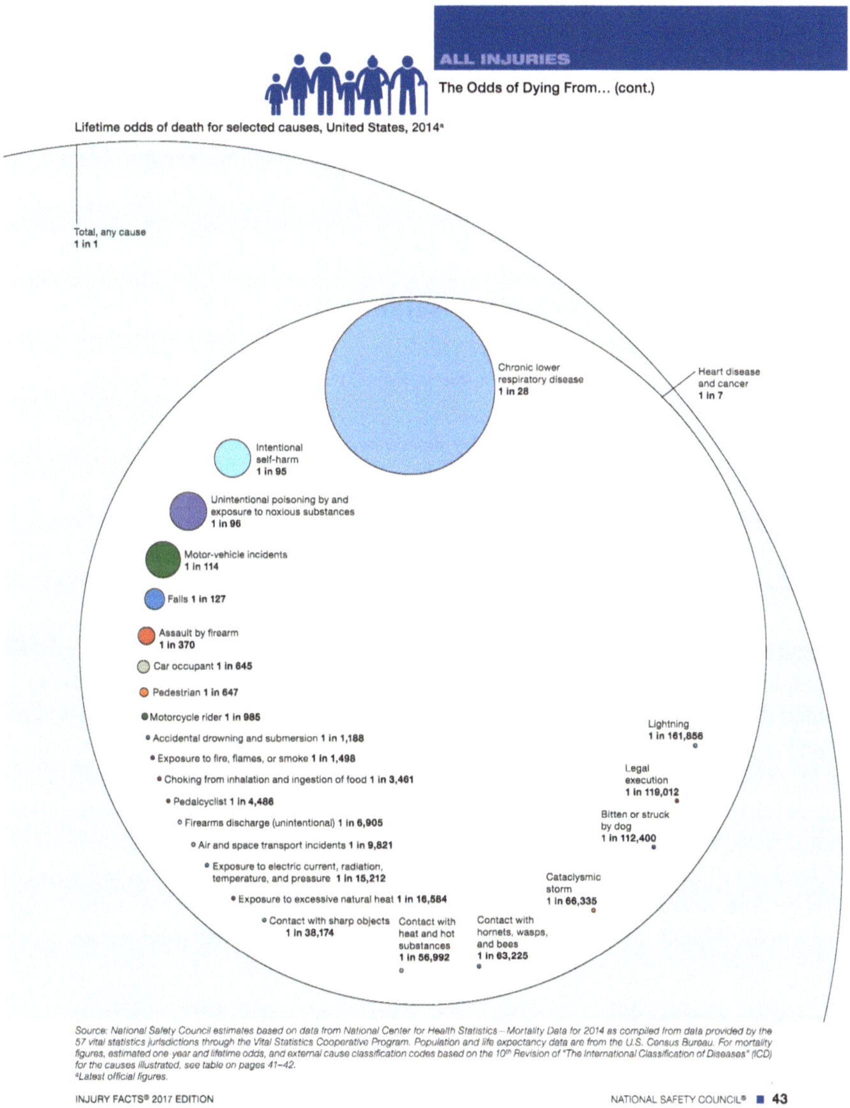


**Fig. 8.2** Lifetime odds of death for selected causes

and *feelings* about safety is central to understanding laypersons' perceptions of aviation safety and the actions they take.

Of course, aviation professionals realize that commercial aviation is very safe. For the pilots, managers, regulators, investigators, and researchers who work in this area, the problem is understanding how to think about safety and proposed safety improvements when there are (fortunately) so few accidents. In automobile traffic safety, there is a natural metric. Interventions are successful when they produce a measurable decrease in the accident or injury rate. But in aviation safety, the accident rate is so low for commercial aviation that it is practically impossible to measure differences in the accident rates between airlines or within the same airline over time. Hence, for aviation professionals, safety is not simply the absence of injuries or accidents. Success must be measured in terms of the impact of safety interventions on the *risk* of accidents. In the sections that follow, the similarities and differences between lay and professional perceptions of aviation safety are explored.

## 8.2 Popular Perception of Aviation Safety

It's important not to focus so much on the statistics, but [on people's] perceptions.

*—Federico Peña, U.S. Transportation Secretary, quoted in* USA Today *newspaper, 22 December 1994.*

### *8.2.1 Perceptions of Risk and Safety*

In general, people have a rough idea of the risk associated with specific hazards. For example, Lichtenstein et al. (1978) asked participants to estimate the number of annual deaths due to 40 causes. The median correlation between the estimated and actual annual number of deaths was .77 in one sample and .66 in another. Slovic et al. (1979) found correlations between .55 and .60 between the perceived and actual ranks of the number of fatalities caused by 30 hazards. These values indicate that there is a moderately strong relation between the estimated and actual number of fatalities, but the association is far from perfect.[2] For example, only approximately 20% of the people participating in Lichtenstein et al. (1978) study thought that stroke caused more deaths than accidents although at the time stroke caused

[2]Correlations indicate the strength of the linear relation between two sets of measures. When the values correspond perfectly (e.g., higher one measure, higher on the other; lower on one measure, lower on the other), the correlation is 1.0. When the values are inversely related (e.g., higher on one measure, lower on the other), the correlation is −1.0. When there is no relation between the measures, the correlation is 0.

85% more deaths than all accidents combined. Death by lightning was judged to be less likely than death by botulism although it was 52 times more frequent. Death by asthma was judged only slightly more frequent than death by botulism, although it was over 900 times more frequent.

People appear to have somewhat more accurate ideas about the safety of commercial aviation than about hazards in general. For example, when Slovic et al. (1979, 2000) asked experts and three groups of non-expert participants to rate the risk of 30 activities, the four groups ranked the risk of commercial aviation nearly identically (16, 16, 17, and 18 respectively). The risk of commercial aviation was ranked similarly to that of riding a bicycle and using electric power. Elvik and Bjørnskau (2005) asked Norwegians to rate the safety of travel by airplane, train, ship, bus, car, motorcycle, bicycle and walking. Correlations between perceived safety and statistical estimates of fatality rates ranged between .75 and .89 depending on the scales used. The great majority of participants (94.7%) perceived commercial aviation to be "safe" or "very safe." Only riding the bus (97.8%) and taking the train (95.1%) were perceived to be "safe" or "very safe" by a larger number of participants.

However, as noted previously, although people generally report believing that commercial aviation is relatively safe, they also often report being anxious about flying. Almost half of the population reports some degree of fear of flying – from slight discomfort to a very intense fear (Capafons et al. 1999). Apparently, a substantial number of people are simultaneously aware that commercial aviation is very safe and somewhat afraid to fly.

These results suggest that different factors affect people's cognitive perceptions of safety and affective feelings of safety. The partial independence between largely affective "System 1" feeling and largely cognitive and analytical "System 2" thinking has been observed in multiple domains (e.g., Zajonc 1980; Kahneman 2011). This duality influences perceived risk in many ways. Risk is not just the perceived likelihood of death. For example, individuals' perceptions of "risk" are only moderately correlated with fatality estimates. Slovic et al. (1979) found that laypersons' risk perceptions across a number of causes correlated between .50 and .62 with annual frequencies of death whereas experts' assessments of risk were correlated .92 with annual frequencies of death. This is not due solely to laypeople misestimating the actual frequencies. If the differences in correlations between laypeople and the experts were due merely to laypersons misestimating the actual frequencies, then the correlations between laypersons' annual fatality estimates and their own perceptions of risk should approximate those of the experts. However, these correlations were only moderate (.47 for one group and .56 for another). Factors other than fatality rates influence lay risk assessments. Some of the discrepancy between laypeople and experts in the correlations noted above can be explained by taking individuals' assessments of the number of fatalities that *could* occur in a particularly bad year into account. Logically, the effect of particularly bad accidents on annual death estimates should have been taken into account in arriving at the original fatality estimates. So, there should be no additional effect of taking catastrophes into account. But there is (Slovic 2000). People are particularly sensitive to disasters. Activities that can produce large numbers of deaths in a single incident are seen as riskier than activities that generally affect a small number of individuals at a time.

In general, individuals perceive hazards to pose greater risk when: (1) they have little control over exposure to the hazard or to its effects, (2) the hazard is personally relevant, (3) the hazard is potentially catastrophic, (4) the effects of the hazard may be unobservable, delayed, or unknown, and (5) the effects of the hazard are "dreaded" (Slovic et al. 1980, 2000). These are factors that draw attention to the activity and engender emotional responses. For example, fear arises when individuals appraise a situation as one in which something unpleasant could happen to them, but the outcome is uncertain, and they may not be able to cope with the effects (e.g., Roseman 1984; Smith and Ellsworth 1985).

Commercial aviation shares many of the characteristics noted above that increase the perception of risk. Compared to drivers of automobiles, passengers in commercial airliners have little or no control over their fate. While driving, one can at least manipulate the controls and maintain the illusion of complete control. In the back of an airplane, even the illusion of control is impossible. Although the passenger in an automobile has much less control than the driver, the proximity to the driver and the familiarity of the surroundings is likely to provide the passenger with some illusion of control. As a passenger in an automobile, one can see the actions of the driver. The movements, the environment, the sensations are all very familiar. Not so for the passengers in commercial aircraft. Furthermore, though aircraft are now commonplace throughout much of the developed world, they remain somewhat mysterious. People typically have only a vague idea of how airplanes fly or what technology is used to fly them. They may not know how wings enable the aircraft to fly or why airplanes have tails. The displays and controls glanced through the open cockpit door during boarding appear complex and inscrutable. Airline passengers *know* that they have very little control over their fates.

In addition, unlike fatal automobile accidents, fatal commercial airline accidents almost always qualify as disasters. Although many more people die in automobile accidents (by several orders of magnitude) than die in commercial aviation accidents, when there is a fatal airliner accident many more people could die in that single incident than in any single traffic accident. Finally, people may find the image of dying in an aircraft accident to be more "dreadful" than dying in an automobile accident. The thought of falling out of control before dying in a fiery explosion is horrendous.

Commercial airline travel also involves other aspects that are emotionally evocative. People travel for a reason. The anticipation of a vacation, reunion, or important business meeting is emotionally arousing. So are the hassles involved with getting to the airport and waiting in lines. All of this subtle emotional excitation can transfer (Zillmann and Bryant 1974) to aspects of the flight itself. Sitting close together with other people is itself emotionally arousing (Zajonc 1965). The associated physiological arousal can amplify minor affective responses as can the emotional contagion (Hatfield et al. 1994) shared between passengers. Put together, these effects can take a little shaking of the airplane in light turbulence and magnify it into a frightening event. When later recalled, these emotional memories can influence perceptions of safety independently of the factual information that people may possess. Thus, the same individuals can recall statistics that indicate that commercial air travel is safe while simultaneously possessing fearful memories and images that lead them to be afraid and to act on those feelings.

### *8.2.2 Effects of Perceptions of Aviation Safety on Economic Behavior*

The effects of feelings about aviation safety can be seen in a variety of behaviors. Perhaps the most direct is the purchase of commercial flight life insurance. Not to be confused with travel insurance, which compensates travelers for trip cancellation, travel delays, lost luggage, etc., flight life insurance pays beneficiaries in the event of the traveler's death. At one time, coin operated flight life insurance machines were found near many airport ticket counters. These have mostly disappeared, replaced by Internet sales. But why do these policies exist? Most life insurance plans cover death due to airline accidents, so why should anyone purchase additional insurance only for one of the least likely hazards? If you were concerned with providing for dependents, then you would purchase standard life insurance that would cover a host of different potential causes of death including airline accidents. If you were not concerned enough to purchase life insurance that covered more likely causes of death, why purchase flight life insurance? One possibility is that it makes you feel better. Based on both laboratory and field experiments, several researchers (Tykocinski 2008; Van Wolferen et al. 2013) have observed that individuals who purchase insurance (including traveler's insurance) or are reminded that they have insurance believe that they are less likely to suffer the insured event than are individuals who did not purchase and/or were not reminded of their insurance. Of course, having insurance can lessen the financial burden of an event but it cannot change its probability. However, it can reduce the feeling of risk.

The effects of feelings about aviation safety also can be seen in individuals' choices of airlines, routes, and flights. In making these choices, price is always important. However, most people demonstrate other preferences as well. For example, most people prefer to fly non-stop and on an airline based in their country. These biases are exaggerated in individuals who express some fear of flying. For example, Fleisher et al. (2012) asked 335 Israeli college students about their responses to flying. A large percentage of the respondents expressed some anxiety associated with flying. Forty-two percent reported that they felt afraid during take-off and landing; 39% felt insecure in the air; 33% reported that "airpockets" made them nervous. The participants were then asked to choose particular airlines and flights for several international routes. Individuals high in fear of flying ($85^{th}$ percentile) were willing to pay substantially more ($66) than individuals low in fear of flying (15$^{th}$ percentile) to fly on their "home" carrier. On one route, individuals high in fear of flying were willing to pay $155 more to fly on their home carrier rather than on an unscheduled foreign airline. Individuals low in fear of flying were willing only to pay $50 more. On another route, individuals who were high in fear of flying were willing to pay $558 more to fly on a scheduled home carrier and $389 more to fly on an unscheduled carrier from their home country than to fly on a foreign carrier. Individuals who were low in fear of flying were willing to pay only $114 more to fly on the home scheduled carrier and $110 more to fly on the home unscheduled carrier. Individuals high in fear of flying were willing to pay $479 more to fly non-stop compared to individuals low in fear of flying who were willing to pay only $181 more.

When an accident happens, the immediate costs can be substantial. The airline may need to pay to replace the aircraft and for damages on the ground. It may also need to make payments to survivors, estates of passengers and crew, companies shipping cargo, etc. Furthermore, the company likely will spend years defending lawsuits. However, the secondary "social amplification" (Kasperson et al. 1988) costs of airline accidents can dwarf the direct costs. For example, following a serious airline accident, the number of tickets purchased for the accident airline decreases (Squalli and Saad 2006; Mitchell and Maloney 1989) resulting in an estimated average loss of over $300 million in ticket revenue for the airline. The airline's stock prices also decline (Borenstein and Zimmerman 1988). Indeed, some airlines have gone out of business following a major accident.

When an airline accident results in limited or no loss of life, there is some evidence that potential travelers shift to other airlines (Ho et al. 2013). However, when many (10 or more) lives are lost, individuals may choose not to travel or to use other means of transportation (Ho et al. 2013). Of course, rationally, this should not occur. People know that commercial aviation is very safe, and they also know that accidents occur. Hence, passengers and investors should have taken the potential for accidents into account when making their decisions. However, the evidence demonstrates that the feelings engendered by an accident have a substantial impact nonetheless.

On the surface, there appears to be no rational reason why accidents with relatively large loss of life should implicate the industry as a whole while accidents with fewer fatalities should implicate only the airline involved. This result could be due, in part, to a bias against regional airlines, which fly smaller aircraft. These airlines may be perceived as not as professional or as safe as the mainline airlines that fly larger aircraft. Thus, if a regional airline has an accident, the event may be taken as confirmation that the particular airline involved in the crash is not safe. Whereas if a mainline airline has an accident, this may be regarded as a signal (Slovic 1987) that the entire industry is not as safe as it appears. Thus, the whole industry suffers. However, the number of fatalities resulting from a crash is not perfectly related to the size of the aircraft. Large and small aircraft can be involved in accidents that result in few if any serious injuries or fatalities. Hence, other factors may be at work. In particular, the spectacle of a large-scale aviation disaster is likely to draw more media attention (Coombs and Slovic 1979; Cobb and Primo

2003) than smaller events, thus extending and enhancing the emotional response to the event. The façade of complete safety (encouraged by airlines and regulators) breaks. The event shocks not only individuals' confidence in the safety of commercial airline travel but in the orderliness and predictability of life. People are shocked and saddened. These emotions translate into reduced activity including economic activity. In the days immediately following major airline accidents, general stock values – not just airline stocks – fall across the board with an estimated average loss of $60 billion per aviation disaster (Kaplanski and Levy 2010).

## 8.3 Perception of Safety Within the Aviation Industry

When we say an airline is safe to fly, it is safe to fly. There is no gray area.

*—David R. Hinson, Federal Aviation Administrator, under oath to the Senate Commerce Committee, 14 May, 1996.*

Pilots, like the general public, believe that airline travel is safe. For example, Chinese pilots rated the risk of flying as being approximately the same as the risks found in "everyday life" (You et al. 2013). In another study, 73% of flight instructors (one of the more dangerous jobs in aviation) reported that they regarded flying as a safe activity. Indeed, with increased experience, pilots appear to perceive less risk in aviation (Thompson et al. 2004). It is not surprising that commercial pilots do not perceive their profession to be particularly dangerous. After all, they chose the profession and know a great deal about it. However, there are substantial differences between the general public and aviation professionals in the *way* that they think about safety.

These conceptual differences have their roots in two fundamental perceptual differences. First, the general public sees safety as a dichotomy; an activity is safe or unsafe. Aviation professionals (like other individuals engaged in highly reliable professions) see safety as a continuum; safety varies in degree. Second, the public sees safety as the absence of harm, the lack of an accident. Aviation professionals see safety as a process, an integrated series of decisions and actions that lower the risk of harm. These differences influence how safety decisions are framed, the information that is considered, and the choices that are made. These differences and their effects are explored in the sections below.

## 8.4 Safety as a Continuum

Safety is often treated as a dichotomy. The question "Is it safe?" is meant to be answered "Yes" or "No." Although treating safety as a dichotomy is more common in public discourse than in professional circles, this tendency is not absent from professional discussions. In aviation, the dichotomization of safety is abetted by airlines and government agencies. Airline managers frequently state that all of their operations are safe. Regulators pronounce that despite obvious differences, all airlines operate at "one level of safety." This labeling has consequences. If your airline is "safe," how can a manager justify additional expenditures on "safety?" If all airlines are "safe," how can a regulator justify imposing additional "safety" regulations? Furthermore, treating safety as a dichotomy leads to public demands that cannot be met. The system cannot attain "absolute safety" with no accidents. So, when an accident occurs, there is a sense of betrayal. If the system is safe, then why did the accident occur? Who is to blame?

Thinking of safety as a matter of degree rather than as a dichotomy can help. Typically, businesses justify expenditures based on the financial "return on investment." This rationale also can be applied to non-monetary benefits. For example, a government traffic engineering department can justify roadway improvements on the basis of the projected reduction in the number of automobile accidents and lives saved. However, in commercial aviation, it is difficult to make this case. Because accident rates are so low, there is rarely any *observable* effect of any single intervention on the accident or fatality rate. This tends to work against investments in safety. On the one hand, if a manager decides to fund a safety intervention and it is successful, nothing will happen. There will be no accidents and no direct way to justify the expenditure. On the other hand, if an accident occurs, the expenditure could appear useless. However, if safety is conceptualized as a process rather than an event, then measuring the components of that process can be used as indicators of safety.

## 8.5 Safety as a Process

The remarkable safety record of commercial aviation in the western world is the result of a system that has evolved to be resistant to mistakes and failures. Nobody is perfect. Equipment fails; people make mistakes. Multiple errors are made on every flight. But there are multiple barriers in place to catch and rectify these problems. For example, a typical procedure for complying with Air Traffic Control (ATC) instructions is for the pilot tasked with handling ATC communication to listen to the controller and write down the instructions. He or she then reads the instructions back to the controller who verifies that they were correctly transmitted. The pilot then enters these into the aircraft's flight management computer at which point the other pilot verifies that the paper copy has been correctly translated into computer instructions. Then, the pilots monitor different displays to verify that the instructions are correctly executed. At the same time, the air traffic controller is monitoring the progress of the flight. In addition, should the aircraft stray too close to terrain or another aircraft, automated warnings onboard the aircraft and in the air traffic control center will be activated. Accidents occur only when a human error is made, or a piece of technology fails, and the system fails to catch it. James Reason (1990) has captured the essence of the safe system in the "Swiss cheese" model (see

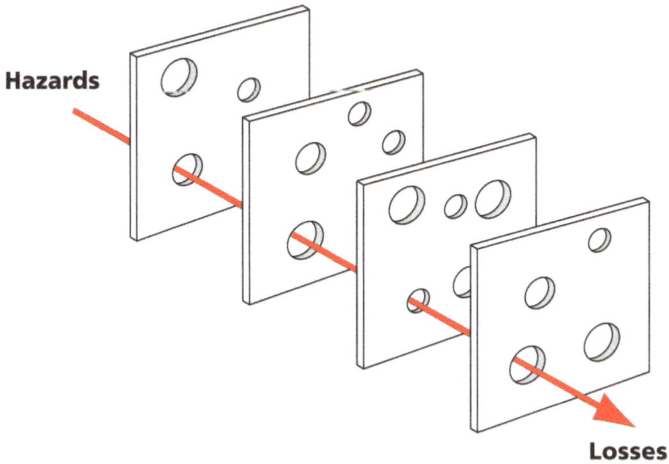


**Fig. 8.3** Reason's (1990) "Swiss cheese" model of hazards

also Proske, Chap. 2) depicted in Fig. 8.3. In this model, an accident occurs *only* when *all* of the barriers fail simultaneously (holes in the cheese align).

From this perspective, safety can be considered to be a product of the components of the system working together and hence measures of the function of those components provides a measure of safety. For example, flying an airplane safely requires that the crew perform a variety of tasks. To ensure that these tasks are performed correctly, airlines adopt well-defined procedures (which are approved by regulators) that determine how crews are to perform the required tasks. Often, these procedures include checklists that the crew must use to verify that the necessary tasks have been performed and that the desired outcomes have been achieved. Therefore, the failure to perform a step in a procedure (including any included checklist) as specified is considered a safety problem whether or not it led to any operationally relevant problem. Of course, adherence to procedures is not the only factor that impacts safety. Other aspects of the operation including maintenance, ground handling, training, crew communication, and the "safety culture" of the airline can impact safety and measures of these may be used as indicators of the safety of the operation.

All airlines in the western world have a safety department (which generally operates according to an approved Safety Management System; ICAO 2013), that monitors operations for signals that things are not as safe as they were presumed to be and prepares for anticipated changes. Major airline safety departments monitor a host of measures that are believed to be indicators of the proper functioning of the components of the system. Problems in these components are perceived to be links in the chains of events that lead to accidents. These may include minor errors in the execution of standard flight or maintenance procedures, equipment problems, navigation issues, etc. Reports of problems may be generated by pilots, maintenance technicians, or other personnel or by automated electronic monitoring systems. By themselves, few if any of these events could lead directly to an accident. In almost every case, the problem is caught by one of the barriers built into the system. But the report is an indication that there may be a problem with a part of the system that may place the operation at greater risk.

## 8.6 Safety as Risk Reduction

A major airline operation is so complex, involving so many components in complex interactions between people, technology, and the physical and social environment (Mauro et al. 2012) that the process appears to be stochastic. Hence, it is appropriate to think in terms of the risk posed by an operation. Risk is a function of likelihood and consequences: the more likely a hazardous outcome is to occur, the greater the risk; the greater the damage that may occur, the greater the risk. Decisions can be framed in terms of the risks attached to the options under consideration. The increase or reduction in risk that would result from each option can be weighed with the expected costs and benefits in choosing between options.

The expected reduction in risk can be used to justify safety investments. However, for most proposed interventions, a concrete expenditure must be weighed against a probabilistic quantity. This is a difficult decision to make and justify. To make sensible safety decisions, pilots, managers, and regulators must have a reasonably accurate perception of the current safety risks and a reasonably accurate assessment of how these risks would change with changes to airline operations.

As in most industries, risk assessments in aviation are typically loosely structured. Often, the task is conceived as requiring that the activity under discussion be evaluated on two dimensions, likelihood of occurrence and severity of consequence, both rated on ordinal scales. The result is represented as a point on a "risk matrix" (see Fig. 8.4). In some cases, a more structured approach may be taken including the use of probabilistic risk analysis (e.g., Bier and Cox 2007). However, it is rare to have all of the data that one would need to calculate the precise probabilities needed to justify the time, effort, and expense required to use these techniques. So, knowledgeable people are gathered, the issues are discussed, and a global rating of the likelihood and severity of the possible consequences is determined by consensus. However, this loosely structured approach is susceptible to a variety of problems that can result in misleading assessments. These problems include difficulties in enumerating the possible outcomes, estimating the likelihood of these outcomes, and determining the potential severity of the consequences. Each of these issues will be discussed below.

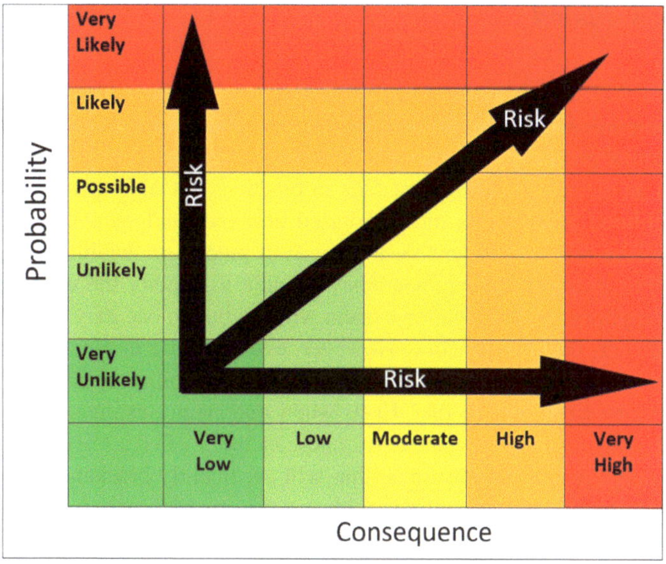


**Fig. 8.4** A typical risk matrix. Greater risk is associated with more serious consequences and higher probabilities of those consequences

### *8.6.1 Possible Outcomes*

To accurately assess the risk posed by a potential problem, one must first consider the possible outcomes that could result if the problem were to occur. Often, individuals attempt to simplify this task by considering only the worst case. This can be misleading. For example, an airline safety analyst might be concerned about a crew's report of anomalies in the operation of the airline's weight and balance calculation program. In the worst case, the aircraft could depart out of balance and encounter an event that causes the aircraft to enter an aerodynamic stall from which recovery is impossible given the weight distribution. However, this is an exceedingly unlikely scenario. A manager might reasonably conclude that this possibility is so remote, that other problems have a higher priority. However, there is a much more likely outcome that should be considered. Aircraft that are improperly loaded may burn substantially more fuel because of the out-of-balance condition. Although this outcome is not catastrophic, over a large number of flights the cost of the error could be large enough to cause substantial financial damage to the airline. It is also not sufficient to consider only the most likely outcome. In many cases, unlikely outcomes have sufficiently serious consequences and are likely enough to be cause for concern. Generating lists of possible outcomes requires domain knowledge and insight. However, in many cases, one can generate outcomes by systematically considering the general classes of factors that are likely to affect the result of a failure or other problem.

### 8.6.2 Likelihood

To proceed with a risk assessment, one must estimate how likely it is that each possible outcome will occur. Sometimes, the probability of a given outcome can be estimated quite precisely. For example, one may have engineering data that indicate how often an electronic component fails in practice. But often this is not the case. Many likelihood assessments must be based on expert judgments. In many cases, experts will be reluctant or unable to specify a precise probability for a possible outcome. For example, an engineer may be able to specify the conditions under which a component of a navigation system will fail but no one may know how often those conditions occur in practice. However, even in these instances, it is rare that one knows nothing. It is rarely the case that the probability of an outcome could plausibly range from zero to one. Even when one cannot estimate the probability associated with an outcome precisely, one can often offer a "best estimate" and specify a range around that estimate that will confidently bracket the actual probability. This is sufficient to continue with the risk assessment.

### 8.6.3 Consequences

Because risk is a function of likelihood and consequence, the possible damage that could result from an event must be assessed. In some cases, assessing the consequences of possible outcomes is relatively straightforward. For example, a manufacturer of desktop computer displays may be concerned about the consequences of failures of the display during the warranty period. In this case, it is relatively easy to assess the possible consequences because only one type of consequence, monetary loss from replacing the display and the labor required to replace it, need be considered. However, in aviation an event could cause many different types of consequences that are not easily measured on a single scale. An event could cause property damage, injury and loss of life, and disrupt operations. Furthermore, an event could generate secondary damage through people's reactions to the original event.

People often attempt to simplify the assessment process by trying to use one measure to scale all of the different types of damage. However, disagreements about the validity of an assessment may arise not because of any debate over the possible consequences or their likelihood but only over the value attached to the consequences. For example, insurance companies and international agreements specify how much the loss of a limb or the death of an airline passenger is worth in dollars. These amounts can then be combined together with estimates of the costs of property damage and lost revenues to arrive at a single monetary value that can be used as the measure of the consequences of an accident. However, people may reasonably disagree with the value attached to life by an insurance company; courts often do. Furthermore, these calculations may lead decision-makers to make trade-offs that they themselves find unacceptable. For example, if an arbitrary monetary value is attached to the value of a life, then the rational decision is to forgo safety investments whenever the costs of those investments exceeds the monetary value of the lives likely to be lost if the investment is not made. Once a monetary value for a life is accepted, the trade-off appears rational although the individuals making the decision may not agree that the value of a life can be reduced to the specified amount.

Risk assessors also often ignore or grossly underestimate the secondary damage caused by people's reactions to an event. For example, the damage caused by a fatal crash of an airliner includes the value of the aircraft, the damage to life, limb, and property in the aircraft and on the ground, and the loss of revenue caused by the loss of the aircraft and the disruption to the schedule. However, the damage caused by the crash also includes the psychological trauma endured by survivors and relatives, increases in fears of flying, and damage to the reputation of the airline and the industry. Some of these costs are borne by the airline or its insurers directly in payments to individuals or indirectly in lost ticket sales and decreased stock values. Some of these costs are borne by the industry in decreased travel and calls for increased governmental oversight. Some of the costs are borne by the society as a whole. In many cases, the costs associated with this "social amplification" (Kasperson et al. 1988) can substantially outweigh all other consequences.

Rather than using a single "severity" dimension, it is often better to evaluate the consequences of an event on separate dimensions that are combined only when general agreement on the combination rules can be established. These dimensions may differ by domain. For commercial aviation, threats to life and health, financial costs, threats to mission success (e.g., arriving at the destination on time), and social amplification may be most useful.

### 8.6.4 Extreme Risks

Assessing outcomes with extreme consequences pose a particularly difficult problem (Kunreuther 2002). In most cases, the traditional calculation of risk as the product of the probability of an event and the potential consequences appears to approximate our sense of what risk is. For example, a business is likely to treat a high likelihood of a small monetary loss as of roughly equivalent risk to a low likelihood of a somewhat larger loss. However, when the probabilities and/or consequences approach their extremes, as they may in commercial aviation, the risk estimate produced by the traditional calculation departs from our intuitions. In particular, an event that could cause a catastrophe with very low probability is generally perceived to be much riskier than an event that is highly likely to cause an outcome with very low cost (Hohenemser et al. 1983).

This is not solely a perceptual phenomenon. Extreme consequences are objectively different. For example, an airline can plan for how to respond physically and financially to most potential outcomes. But one cannot plan for how to respond if the consequence is the collapse of the company. Once the company is out of business, nothing can be done. There is a discontinuity in the risk function at the point at which the consequences become unbearable. One cannot treat the collapse of the company, the destruction of an ecosystem, or the death of a society as simply an outcome with very high costs. This does not mean that one cannot assess extreme risks, only that one should not rely on the mechanical application of any simple risk calculation procedure in all situations.

The common unstructured approach to risk assessment is in part meant to avoid System 1 thinking and encourage individuals to make "rational" safety decisions. However, as discussed above, the lack of structure creates opportunities for multiple problems to emerge. Although following a more structured approach (e.g., Mauro and Barshi 2009) will not eliminate these problems entirely, it can limit them. A structured approach to risk assessment can alter the focus from producing a unitary assessment of risk to the particular steps involved – determining the possible outcomes, assessing the likelihoods of these outcomes, and the severity of consequences. This focuses the discussion on likely sources of disagreement (e.g., estimates of likelihood, severity of consequences) rather than on a vague overall assessment of risk.

## 8.7 Conclusion

> If one took no chances, one would not fly at all. Safety lies in the judgment of the chances one takes.
>
> —*Charles Lindbergh, journal entry 26 August 1938, published in* The Wartime Journals, *1970.*

Aviation professionals and the general population have somewhat different perceptions of aviation safety. In the general population, there is a conflict between facts and fears. People know that commercial aviation in the western world is extremely safe, but they do not always feel safe and often make decisions based on "System 1" feeling rather than "System 2" thinking. For aviation professionals, the conflict is between facts and facts, with managers and regulators struggling to weigh costs and benefits with risks in making safety related decisions.

In the general population, people regard safety as the absence of accidents. Because of the extremely low accident rate in commercial aviation, this makes it impossible to compare most airlines, aircraft, operations, routes, etc. on safety. Aviation professionals regard safety as a process and use measures of the components of that process as indicators of the safety of the system. This allows meaningful comparisons to be made even with the very small number of accidents.

In the general population, people ask "is it safe?" They are frequently reassured by airlines and regulators that flying is safe. But no flight operation is without some risk. When an accident occurs, people are shocked, horrified, and betrayed by those that guaranteed their safety. The first targets are frequently the pilots. When the pilots fail to take the steps required to counter the effects of problems in equipment design, maintenance, procedure design, or training the accident is blamed on "pilot error." Indeed, in some countries, pilots that survive may be held legally responsible for accidents that are not the result of intentional malfeasance. But the airline or regulators may be blamed as well for failing to adequately protect the public. Political pressures may lead to counterproductive rules or legislation that may not be cost effective or even effective.

Explicitly recognizing that there is a continuum of risk and degrees of safety would encourage intelligent public and professional discourse. Regulators and airline managers would be encouraged to make operations safer by taking intelligent steps to increase safety based on the assessed risk. There is nothing that will ease the pain of the friends and families of victims, but perhaps the sense of betrayal would be lessened if we admitted that there was always some risk. However, thinking in terms of risk with the attendant focus on probability would require a dramatic societal change. Western cultures support a deterministic view of the world. This supports a "can-do" attitude but leads to rejection of complex partial solutions for complex problems. At the same time, large segments of many populations are ignorant of the basic concepts of probability while even educated individuals often have difficulty applying those concepts outside of the context in which they were learned. Hence, attempting to change attitudes towards safety and risk likely would be very difficult.

Aviation professionals ask not whether "it is safe" but rather "how safe is it?" For aviation professionals, the challenge is making accurate risk assessments. The ultimate purpose of performing risk assessments is to inform decisions about potential actions. When a safety concern is raised, airline managers must decide whether the risk warrants an intervention. Managers must weigh the differences in potential risk reduction of different courses of action against the differences in cost. Regulators must make parallel analyses in determining whether to allow airlines to make particular changes or in mandating implementation of particular safety interventions. There may be costs that are too high and levels of risk that are unacceptable. However, in most cases, the question is whether the reduction in risk is worth the investment. A careful risk assessment can clarify these decisions. However, at some point the airline, the regulators, and the society as a whole must ask, "how safe is safe enough?" All risk cannot be eliminated. Safety interventions cost money. These costs must be borne by the airlines, the passengers, and the society as a whole, and the money spent on these interventions cannot be spent elsewhere.

Is the present level of risk acceptable? How safe is safe enough? These questions cannot be answered in the abstract but only by weighing the risks, costs, and benefits of alternative courses of action (Fischhoff et al. 1981). If by 2025, the amount of commercial air traffic were to increase as expected (FAA 2007) and the present accident rate held constant, the number of accidents could increase to as many as one accident per week. Would this outcome be acceptable, or should we invest additional resources and try to reduce the risk to maintain the same number of accidents? The answer may be different if you are a pilot, crewmember, or passenger, but it's a question that must be asked and answered.

## References

Bier, V. M., & Cox, L. A. (2007). Probabilistic risk analysis for engineered systems. In W. Edwards, R. F. Miles, & D. von Winterfeldt (Eds.), *Advances in decision analysis: From foundations to applications* (pp. 279–301). New York: Cambridge University Press.

Borenstein, S., & Zimmerman, M. B. (1988). Market incentives for safe commercial airline operation. *American Economic Review, 78*(5), 913–935.

Capafons, J., Sosa, C., & Vina, C. (1999). A reattributional training program as a therapeutic strategy for fear of flying. *Journal of Behavior Therapy and Experimental Psychiatry, 30,* 259–272.

Cobb, R., & Primo, D. (2003). *The plane truth: airline crashes, the media, and transportation policy*. Brookings Institution: Washington D.C.

Coombs, B., & Slovic, P. (1979). Newspaper coverage of causes of death. *Journalism Quarterly, 56*(837–843), 849.

CBS (2014). CBS News Poll conducted March 20–23, 2014. Retrieved from PollingReport.com.

Elvik, R., & Bjørnskau, T. (2005). How accurately does the public perceive differences in transport risks? An exploratory analysis of scales representing perceived risk. *Accident Analysis and Prevention, 37,* 1005–1011.

Federal Aviation Administration (FAA) (2007). Capacity Needs in the National Airspace System 2007–2025. Federal Aviation Administration.

Fischhoff, B., Lichtenstein, S., Slovic, P., Derby, S., & Keeney, R. (1981). *Acceptable Risk*. New York: Cambridge University Press.

Fleisher, A., Tchetchik, A., & Toledo, T. (2012). The impact of fear on travelers' flight choice: Choice model with latent variables. *Journal of Travel Research, 51,* 653–663.

Frankovic, K. (2015). Most Americans now worried about flying. Economist/YouGov poll. https://today.yougov.com/news/2015/04/01/most-americans-now-worried-about-flying.

Gallup (2006). Poll conducted August 18–20, 2006. Gallup2 http://www.gallup.com/poll/1579/airline).a)px.

Hatfield, E., Cacioppo, J., & Rapson, R. (1994). *Emotional contagion*. New York: Cambridge University Press.

Ho, J., Qui, M., & Tang, X. (2013). Do airlines always suffer from crashes? *Economics Letters, 118,* 113–117.

Hohenemser, R., Kates, R., & Slovic, P. (1983). The nature of technological hazard. *Science, 220,* 378–384.

International Air Transport Association (IATA) (2015). Economic Performance of the Airline Industry.

International Civil Aviation Organization (ICAO). (2013). *Safety management system manual* (3rd ed.). ICAO: Montreal.

Ipsos-McClatchly (2010). Poll conducted by Ipsos Public Affairs January 7–11, 2010. Retrieved from PollingReport.com.

Kahneman, D. (2011). *Thinking, fast and slow*. New York: Farrar, Straus and Giroux.

Kaplanski, G., & Levy, H. (2010). Sentiment and stock prices: The case of aviation disasters. *Journal of Financial Economics, 95,* 174–201.

Kasperson, R. E., Renn, O., Slovic, P., & Brown, H. S. (1988). The social amplification of risk: A conceptual framework. *Risk Analysis, 8,* 177–187.

Kunreuther, H. (2002). Risk analysis and risk management in an uncertain world. *Risk Analysis, 22,* 655–664.

Lichtenstein, S., Slovic, P., Fischhoff, B., Layman, M., & Coombs, B. (1978). Judged frequency of lethal events. *Journal of Experimental Psychology: Human Learning & Memory, 4,* 551–578.

Mauro, R. & Barshi, I. (2009). Risk assessment in aviation. In: Proceedings of the 15th international symposium on aviation psychology, Wright State University.

Mauro, R., Degani, A., Loukopoulos, L., & Barshi, I. (2012). The operational context of procedures and checklists in commercial aviation. In: Proceedings of the human factors & ergonomics society 56th annual meeting, pp. 758–762.

Mitchell, M., & Maloney, M. (1989). Crisis in the cockpit? The role of market forces in promoting air travel safety. *Journal of Law and Economics, 32,* 329–355.

National Safety Council (2015). Injury facts. http://www.nsc.org/learn/safety-knowledge/Pages/injury-facts-chart.aspx.

Reason, J. (1990). *Human error*. New York: Cambridge University Press.

Roseman, I. (1984). Cognitive determinants of emotion: A structural theory. In P. Shaver (Ed.), *Review of personality and social psychology: emotions, relationships, and helath* (Vol. 5, pp. 11–36). Beverly Hills: Sage.

Slovic, P., Fischhoff, B., & Lichtenstein, S. (1979). Rating the risks. *Environment, 2,* 14–20.

Slovic, P., Fischhoff, B., & Lichtenstein, S. (2000). Facts and fears: Understanding perceived risk. In P. Slovic (Ed.), *The perception of risk*. London: EarthScan Publications.

Slovic, P. (2000). *The perception of risk*. London: EarthScan Publications.

Slovic, P. (1987). Perception of risk. *Science, 236,* 280–285.

Smith, C., & Ellsworth, P. (1985). Patterns of cognitive appraisal in emotion. *Journal of Personality and Social Psychology, 48,* 813–838.

Squalli, J., & Saad, M. (2006). Accidents, airline safety perceptions and consumer demand. *Journal of Economics and Finance, 30,* 297–305.

Thompson, M., Onkal, D., Avcioglu, A., & Goodwin, P. (2004). *Risk Analysis, 24,* 1585–1595.

Tykocinski, O. (2008). Insurance, risk, and magical thinking. *Personality and Social Psychology Bulletin, 34,* 1346–1356.

Van Wolferen, J., Inbar, Y., & Zeelenberg, M. (2013). Magical thinking in predictions of negative events: Evidence for tempting fate but not for a protection effect. *Judgment and Decision Making, 8,* 45–54.

You, X., Ming, J., & Han, H. (2013). The effects of risk perception and flight experience on airline pilots' locus of control with regard to safety operation behaviors. *Accident Analysis and Prevention, 57,* 131–139.

Zajonc, R. (1965). Social facilitation. *Science, 149,* 269–274.

Zajonc, R. (1980). Feeling and thinking: Preferences need no inferences. *American Psychologist, 35,* 151–175.

Zillmann, D., & Bryant, J. (1974). Effect of residual excitation on the emotional response to provocation and delayed aggressive behavior. *Journal of Personality and Social Psychology, 30,* 782–791.

# Chapter 9
# Perceived Safety While Engaging in Risk Sports

**Martina Raue, Bernhard Streicher, Eva Lermer and Dieter Frey**

**Abstract** Risk sports such as mountaineering, climbing, or backcountry skiing have become quite popular and more easily accessible for the general population. Easy access and added protection may increase perceived safety and result in underestimations of harm. In this chapter, we discuss the interplay of cognition, emotions and bodily states when judging risks while being physically active. We introduce three field and one laboratory study, in which we investigated the influence of physical activity on risk perception. In the field studies, less experienced participants made lower risk judgments under physical activity than before the activity. In the laboratory study, general health-related risks were judged as lower under physical activity. Thus, our findings indicate that physical activity does affect risk judgements. We discuss practical implications on how to support especially less experienced people in making good judgments when engaging in risky activities.

M. Raue (✉)
AgeLab, Massachusetts Institute of Technology, 77 Massachusetts Avenue, Cambridge, MA, USA
e-mail: maraue@mit.edu

B. Streicher
Institute of Psychology, UMIT—Private University for Health Sciences, Medical Informatics and Technology, Eduard-Wallnöfer-Zentrum 1, 6060 Hall in Tyrol, Austria

E. Lermer
Department of Experimental Psychology, University of Regensburg, Universitätsstr. 31, 93053 Regensburg, Germany

E. Lermer
FOM University of Applied Sciences for Economics and Management, Arnulfstr. 30, 80335 Munich, Germany

D. Frey
Department of Social Psychology, Ludwig Maximilian University Munich, Leopoldstr. 13, 80802 Munich, Germany



M. Raue et al. (eds.), *Perceived Safety*, Risk Engineering,
https://doi.org/10.1007/978-3-030-11456-5_9

**Keywords** Physical activity · Risk perception · Embodiment · Backcountry skiing · Rock climbing

## 9.1 Perceived Safety While Engaging in Risk Sports

The growing popularity of risk sports led to more accessibility and security of areas that used to be reserved for experts. For example, via ferratas have been installed throughout the alps, which are protected high altitude climbing routes along steel cables. However, perceived safety through added protection can be an illusion, because risks in these types of activities cannot be eliminated and excessive risk protection might result in an underestimation of potential harms. The German Alpine Club reported a drastic increase in rescues due to so-called blockages. More and more people have to be rescued from the mountains without having an injury, but rather due to excessive demands (Deutscher Alpenverein 2012). This highlights that people often overestimate their skills and have a false feeling of safety and, as a consequence, underestimate risks when they engage in these kinds of activities. Increasing awareness for influences on perceived safety could be a first step to prevent emergency situations.

Risk perception in sports has received quite some attention in personality psychology. The construct of sensation seeking (Zuckerman 1979), for example, categorizes people according to their individual levels of preferred arousal. In order to elevate or maintain a high level of arousal, sensation seekers participate in activities of high arousal such as risk sports (Zuckerman 2007). However, a high individual tendency to seek risks does not necessarily imply a perception of risks as less harmful. For example, a base jumper may judge the risk of base-jumping as high, but still engages in this activity. Subjective perceptions of risk and safety seem to be more influenced by other factors than personality traits. For example, situational cues such as familiarity with a terrain or consistency with an initial decision have been shown to increase risk taking behavior of backcountry skiers, who constantly have to judge the risk of avalanches (McCammon 2004). In this chapter, we will discuss the influences of experience, intuition, emotions and especially physical activity on perceived safety and risk judgments in high risk sports. We will introduce a series of studies, which suggest that being physically active can decrease risk estimates, especially among those with less experience (Raue et al. 2017, 2018).

## 9.2 Risk as Feeling

Subjective risk perception is usually accompanied by some form of emotion which leads to a reliance on intuition or feelings and often disregards objective facts when judging risks (Slovic et al. 2004). This assumption builds on dual-process theories, which propose two systems of mental processing. While system 1 is intuitive, fast, automatic and often emotionally charged, system 2 is analytic, slow and effortful (Kahneman 2011). Additionally, humans' cognitive capacities underlie boundaries (cf. Simon 1955), because humans can only process a limited number of information at a time. These cognitive boundaries can hinder a careful analysis of all possible options and lead to a shift to more intuitive thinking when judging risks. This is especially the case when one has to make a quick decision under time pressure (Finucane et al. 2000), for example, when cycling in city traffic. Thus, feelings play a major role in risk perception, which has been integrated in the *affect heuristic* and the *risk-as-feelings hypothesis* (Loewenstein et al. 2001). The affect heuristic states that people rely on quick and automatic affective responses that occur when they encounter a risk. The risk-as-feelings hypothesis goes in a similar direction and states that people tend to be driven by the emotional reaction to a risk, which often differs from its cognitive evaluation.

For example, in simple investment choices healthy people lost money, while patients who lacked affect due to neurological impairment did not. Participants did not differ in the first round of investments, but healthy participants adopted a more conservative strategy and invested less in subsequent rounds in reaction to the first outcome. Patients with focal lesions that affected emotional processing were not influenced by the outcomes of previous investment rounds and constantly followed a profit maximization strategy (Shiv et al. 2005). Even though this study did not investigate the type of affect involved (e.g., positive or negative), it highlights the major role of affect in risky decision making. This not only applies to investment decisions; affect influences decision making in many situations that involve risk. For example, when backcountry skiers have to decide whether to ride a steep slope or not (and as a consequence risk to trigger an avalanche), they should carefully analyze the situation and consider objective facts such as the avalanche report, weather and degree of steepness. But, according to the introduced research, we assume that it is more likely that they rely on their feelings in the first place and use these as information to guide their judgment (Peters et al. 2006; Slovic and Weber 2002). Nevertheless, a cognitive analysis in terms of system 2 processing can follow emotional system 1 processing (Bechara and Damasio 2005). As a result, careful cognitive considerations (system 2) can potentially alter emotion-based decision tendencies.

Reliable emotional risk judgments are assumed to mainly rely on prior experience, because intuition develops through experience (Betsch 2008; Klein 1993). In other words, only when a person has encountered similar situations before is the person able to make an adequate decision based on feelings. This does not imply, for example, that experienced mountaineers neglect objective risk factors, but that their initial judgment most likely does not rely on an analytical decision. In risky situations, especially in more complex ones or under time pressure, intuitive decision making is suggested to be superior to analytical thinking due to the general limitation of cognitive capacities. For example, firefighters indicate that they do not carefully consider alternatives when in an emergency situation, but rather make quick intuitive decisions based on experience and constantly adapt these to the situation (Klein et al. 1986). In a mountaineering setting, it was shown that experienced alpine rock climbers are usually able to adequately judge the risk of changes in weather or of reaching the limit of their personal capabilities. However, inexperienced climbers lack this skill and make inadequate risk judgments (Csikszentmihalyi 2000; Delle Fave et al. 2003). Furthermore, a study on avalanche victims highlighted that many of those who had exposed themselves to high risks had completed a basic avalanche training, but lacked practical experience in applying what they had learned (McCammon 2000).

## 9.3 Embodied Risk Perception

Besides experience, we suggest that bodily states are another major influence of subjective risk perception in sports. Very little research exists regarding the influence of bodily states, such as physical activity, on risk assessment and risk-taking behavior. In psychology, the influence of bodily states on cognition and behavior is known as *embodiment* and implies that sensory, motor and perceptual processes influence cognitive, affective and behavioral processes (Wilson and Golonka 2013). Since risk assessments in sports are usually made while being physically active, it is especially important to understand the impact of embodiment on risk perception.

Cognition is not an isolated mental process, but rather all cognitive and affective processes rely on a physical context, including postures and bodily movements (Wilson and Golonka 2013; Wilson 2002). For example, studies on embodiment have shown that people judge others as warmer when holding a warm rather than a cold cup of coffee, or judge a room as being cooler after having been socially rejected (Williams and Bargh 2008; Zhong and Leonardelli 2008). Another experiment showed the influence of bodily states on affective processes. People who held a pen between their teeth in a way that activated the smiling muscular found cartoons funnier than people who held a pen between their lips in a way that was incompatible with smiling (Strack et al. 1988).

However, risk-taking is barely covered in research on embodiment. In one study, it was demonstrated that participants who had experienced a period of bad luck when gambling were more risk-taking after having washed their hands (i.e., washing off bad luck; Xu et al. 2012). Concerning physical activity, a study showed that engaging in excessive competitive exercise (playing tennis) can increase risk-taking in an unrelated task (The Balloon Analog Risk Task, BART) follwing that exercise (Black et al. 2013). The authors offer two explanations for this finding. One explanation states that an increase in the level of dopamine leads to a search for more stimulation through risk-taking. Another explanation states that the increase in risk-taking results from performance errors, which are assumed to be caused by tiredness due to extensive exercising. While moderate exercise can improve cognitive performance, vigorous exercise can decrease performance, resulting in an inverted u-shaped curve that describes the interplay of exercise and cognitive performance. This shape is assumed to result from the dual task effect, indicating that an increase in energy demands shifts the focus from cognitive processing to the control of movements, resulting in a decrease of energy resources for cognitive tasks during extensive exercise (Lambourne and Tomporowski 2010). However, additional explanations, such as changes in the affective state, need to be considered concerning the interplay of physical activity and risk perception.

## 9.4 Arousal and Affect

Research suggests that engaging in exercise releases endorphins, which can reduce anxiety and elevate mood states (Dishman and O'Connor 2009). Mood state, on the other hand, directly influences risk perception and risk-taking behavior. For example, negative mood has been shown to produce an increase in the judged frequency of risks, while positive mood produced a decrease, even if the source of emotion was unrelated to the target (Johnson and Tversky 1983). Furthermore, individuals judge situations as safe or unsafe based on their affective states. A positive affective state signals that a situation is safe, which in turn can increase risk-taking behavior (Yuen and Lee 2003).

However, differing between negative and positive feelings does not exclusively capture the interplay of risk perception and affective state. Hence, the circumplex model of affect (Russell 1980) not only includes the dimension of pleasantness (positive or negative feelings), but also the dimension of arousal (aroused or quiet) in its description of affective reactions (Mano 1992). Research has shown that people under high arousal such as anger (Rydell et al. 2008) or sexual arousal (Ariely and Loewenstein 2006) are more risk-taking than people in a neutral state. Dual-process theories offer an explanation for this link between arousal and risk-taking: higher levels of arousal reduce cognitive capacities and, therefore, interfere with system 2 and can lead to more intuitive thinking (system 1) as well as a disregard of decision-related attributes or cues (Kahneman and Frederick 2002).

Additionally, in risk sports, anxiety is an important emotion, which can limit information processing and thus harm cognitive performance. If cognitive processes are limited due to high anxiety levels, risk judgments based on system 2 may be limited as well. Research has shown that especially anxious, inexperienced people tend to shift their focus too much on worrying about than focusing on a task at hand (Eysenck and Calvo 1992). Per definition, anxiety is a state of stress and high arousal that arises in situations with the potential for an undesirable outcome (Brooks and Schweitzer 2011). There is a bi-directional effect of anxiety and risk perception: not only can anxiety increase perceived risks, but perceiving risks as high can also increase anxiety (see also Beck and Clark 1997).

## 9.5 Physical Activity and Risk Perception

To further investigate the interplay of physical activity and risk perception, we conducted three field studies and one laboratory study (Raue et al. 2017, 2018). In the first field study, people who intended to go backcountry skiing were asked to judge potential risks of this activity at six measurement points before, during and after the tour. Risks were not only perceived as less likely during the activity than before the activity, but risk judgments also constantly decreased during the phase of physical activity. However, this was only the case for less experienced participants. The judgment of participants with higher levels of experience did not change over the six measurement points; they judged the risks as equally before, during and after the tour (see Fig. 9.1).

Avalanche danger level on the survey days ranged from 1 (*low danger level*) to 3 (*considerable danger level*), based on the scale of the European Avalanche Warning Services. Even though more experienced participants were generally more risk seeking concerning the character of the tour (i.e., went on tour during higher levels of avalanche danger), their risk judgments were overall lower than those of less experienced participants. Only at the very last measurement point, after completing the tour, did experience no longer make a difference. At this point, risk perception of less experienced participants dropped to the level of more experienced participants. While risk judgments of more experienced participants depended on avalanche danger level, judgments of less experienced participants did not. This indicates that more experienced participants made more realistic judgments than less experienced participants. Interestingly, the difference between the two groups was largest at the safest conditions (i.e., avalanche danger level 1).

In a second field study, again participants of a risky outdoor sport were approached. We asked people who intended to go on a via ferrata to judge risks related to this activity at five measurement points. Just like backcountry skiing, via ferratas have become very popular in the alps. They offer almost permanent

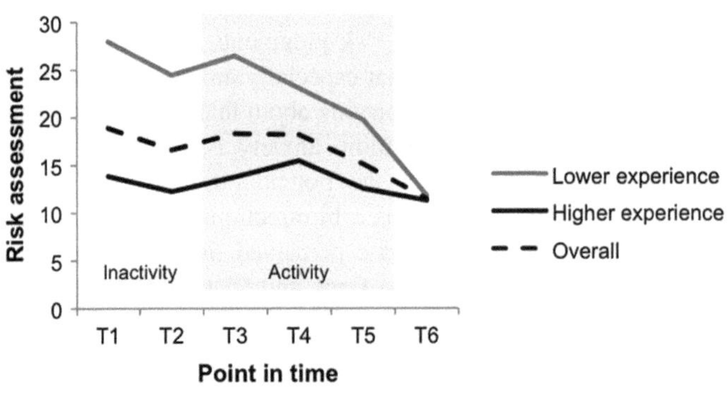


**Fig. 9.1** Risk judgments at different points in time and phases of inactivity versus phases of activity (Raue et al. 2017; see also www.tandfonline.com)

ground-fall-protection through a steel cable, and thus enable even inexperienced people to enjoy alpine settings in a more extreme form. This aspect differs from backcountry skiing, and therefore participants might perceive less threat and more safety. However, accidents also happen on via ferratas, which might be due to an underestimation of the hazards in alpine settings in general, and of the risks of injuries caused by falling on a via ferrata in particular.

Across all measurement points, less experienced participants judged the risks as more likely than more experienced participants. In line with the previous study, risk estimates of less experienced participants tended to be lower during the tour than before the tour, indicating an influence of being physically active. This was especially shown for male participants. However, lower risk perception among less experienced participants before the start of the tour could also be attributed to uncertainty around the unknown. Therefore, a subsequent decrease in risk perception could be the result of more familiarity, physical activity, or a combination of both.

In order to get a better understanding of the findings, we conducted a more controlled third field study in a climbing gym. Similar to via ferratas, climbing in a gym involves constant protection through a rope, but accidents due to human error or gear failure can be fatal. Two important findings emerged from this study. First, participants' perception of climbing risks generally decreased during the activity phase as opposed to the pre-activity phase, independent of their level of experience. Climbing experience only had an influence on the specific risk of falling, which was generally judged as higher by less experienced participants as opposed to more experienced participants. Second, less experienced climbers judged the risk of falling as significantly lower during the activity as opposed to the inactivity phases, but there was no significant change among more experienced participants. The latter finding replicated our results from the preceding field studies, but also ruled out the explanation of familiarity as risk perceptions increased again after the activity phase.

However, the high level of perceived safety due to a permanent fixation to the rock or the gym wall, may have limited some of the findings in the climbing studies. To further investigate the influence of physical activity on risk perception in an even more controlled setting, we conducted a laboratory study. In this laboratory study, participants rode on a stationary bike with medium intensity for two minutes, while a control group sat still at a desk. A repeated measurement was conducted in order to control for individual differences and a control group was included in case of memory effects. Participants filled out an online questionnaire one week before coming to the laboratory. They were asked to estimate the probability of bicycling-related risks (e.g., accident with a car) and bicycling-unrelated risks (e.g., getting a severe illness). We found evidence that physical activity decreases risk estimates, but only concerning general risks and not bicycling-related risks. In this study, the general risks might have been perceived as a greater harm than the bicycling-related risks, and the latter might not have been perceived as a potential risk at all. Thus, although not related to the activity, the results support our previous findings that physical activity can decrease risk estimates. In other words, our findings suggest that people perceive more safety while they are physically active.

In sum, the field studies showed that less experienced participants make lower risk judgments under physical activity than before the activity. The judgments of more experienced participants stayed on a constant level that was below the level of less experienced participants. However, experience did not have an influence in a laboratory study, which investigated bicycling-related and bicycling-unrelated risks while riding on a stationary bike. It was assumed that experience does not play a role in a bicycling setting because it is a highly automated activity that is most likely not perceived to be as dangerous as the activities in the field studies. This might also have been the reason why physical activity decreased risk judgments of bicycling-unrelated risks, but not of bicycling-related risks.

## 9.6 Experience and Risk Perception

The introduced field studies highlight that experience plays a major role when investigating situational influences, such as physical activity, on the assessment of risks. Experience leads to a reliance on intuitive thinking when judging risks, but nevertheless more experienced participants seemed to be able to easily integrate important objective information. In the backcountry skiing study, more experienced participants made more appropriate judgments in relation to the avalanche risk than less experienced participants. Their judgments increased with an increase of avalanche danger level. However, it might also be possible that they simply adhered to their initial judgment and were not open to new information. According to research on the selective exposure to information (Frey 1986), people prefer to integrate information that supports their initial decision and often neglect information that contradicts their initial decision (see also theory of cognitive dissonance; Festinger 1957). Future research should further investigate whether this kind of behavior might be a problem among more experienced participants in risk sports.

Less experienced participants, on the other hand, did not seem to include the avalanche danger level in their judgments and were overall more worried than more experienced participants. This indicates that less experienced people failed to adapt their behavior to objective risk factors. Additionally, being physically active in combination with being anxious might promote extensive arousal, which hinders accurate information processing. A high level of arousal can result in a shift from analytical to more intuitive thinking. However, less experienced people do not have the necessary base of experience when making intuitive judgments. Therefore, this shift is considered especially problematic for this group of people.

In conclusion, more experienced people seem to be less affected by an elevation of arousal because they can rely on their intuitions. However, less experienced people need to analyze the situation and cope with a reduction of analytical skills when the body needs the energy for exercising and the control of movements.

## 9.7 Practical Implications

The findings on less experienced people suggest that this group of people needs to be equipped with simple rules of thumb to compensate for limited experience and support the adequate use of systems 1 and 2. If possible, these rules should be carefully considered before physical activity interferes with accurate information processing. Such simple rules are already used in avalanche trainings (e.g., if the avalanche danger level is 3 or higher, the slope steepness should not increase 30°) and have turned out to be very helpful when making judgments in complex situations. Thus, this approach might be adopted in other risk sports and even in every day decision making such as driving a car (e.g., if it rains, do not drive faster than 100 km/h on the highway). However, if one is very experienced in a situation, such simple rules of thumb might not make sense, but rather might hinder the integration of additional information that is difficult to judge with less experience (e.g., newly paved highway, new tires). Humans are constantly and unconsciously using simple rules of thumb, also known as heuristics (e.g., Tversky and Kahneman 1974; Raue and Scholl 2018; Gigerenzer, Todd, & The ABC Research Group 1999) that often help us overcome our limited cognitive capacities. Given this background on human nature, it makes perfect sense to actively make use of simple heuristics in risky situations, and also to adapt the rules based on experience and cognitive capacity.

In order to cope with interfering emotions such as anxiety when making decisions, research on emotion regulation has suggested different strategies to adapt one's emotions to the actual situation. One such strategy is cognitive reappraisal, which means mentally changing the meaning of a situation in order to decrease its emotional impact (Gross 1998). For example, an upcoming adventure that causes anxiety could be reappraised into a challenge. Research suggests that during physical activity, anxiety may be more easily reappraised into a positive affective state, because arousal and anxiety are congruent concerning the arousal dimension (Brooks 2014). These strategies could also be applied in situations of risk in which strong emotions hinder accurate risk perceptions. Future research is needed, however, to further understand the link between physical activity and perceived safety, and ultimately to suggest appropriate coping strategies.

In conclusion, our findings support earlier research that bodily sensations such as emotions and physical arousal can facilitate good decision-making, but may also lead one astray. Therefore, a major challenge when confronted with risk and uncertainty is to sharpen one's introspection in order to be able to distinguish between helpful and misleading states.

**Acknowledgements** We thank Alexa Balmuth for proofreading this chapter.

# References


Ariely, D., & Loewenstein, G. (2006). The heat of the moment: the effect of sexual arousal on sexual decision making. *Journal of Behavioral Decision Making, 19*(2), 87–98. https://doi.org/10.1002/bdm.501.

Bechara, A., & Damasio, A. R. (2005). The somatic marker hypothesis: a neural theory of economic decision. *Games and Economic Behavior, 52*(2), 336–372.

Beck, A. T., & Clark, D. A. (1997). An information processing model of anxiety: automatic and strategic processes. *Behaviour Research and Therapy, 35*(1), 49–58. https://doi.org/10.1016/S0005-7967(96)00069-1.

Betsch, T. (2008). The nature of intuition and its neglect in research on judgment and decision making. In H. Plessner, C. Betsch, & T. Betsch (Eds.), *Intuition in judgment and decision making* (pp. 3–22). Mahaw, NJ.

Black, A. C., Hochman, E., & Rosen, M. I. (2013). Acute effects of competitive exercise on risk-taking in a sample of adolescent male athletes. *Journal of Applied Sport Psychology, 25* (2), 175–179. https://doi.org/10.1080/10413200.2012.704621.

Brooks, A. W. (2014). Get excited: reappraising pre-performance anxiety as excitement. *Journal of Experimental Psychology: General, 143*(3), 1144–1158. https://doi.org/10.1037/a0035325.

Brooks, A. W., & Schweitzer, M. E. (2011). Can Nervous Nelly negotiate? How anxiety causes negotiators to make low first offers, exit early, and earn less profit. *Organizational Behavior and Human Decision Processes, 115*(1), 43–54. https://doi.org/10.1016/j.obhdp.2011.01.008.

Csikszentmihalyi, M. (2000). *Beyond boredom and anxiety*. San Francisco: Jossey-Bass.

Delle Fave, A., Bassi, M., & Massimini, F. (2003). Quality of experience and risk perception in high-altitude rock climbing. *Journal of Applied Sport Psychology, 15,* 82–99.

Deutscher Alpenverein (2012). DAV-Unfallstatistik 2012. http://www.alpenverein.de/bergsport/sicherheit/bergunfallstatistik/dav-bergunfallstatistik-2012_aid_13110.html. Accessed 20 Sept 2013.

Dishman, R. K., & O'Connor, P. J. (2009). Lessons in exercise neurobiology: the case of endorphins. *Mental Health and Physical Activity, 2*(1), 4–9. https://doi.org/10.1016/j.mhpa.2009.01.002.

Eysenck, M. W., & Calvo, M. G. (1992). Anxiety and performance: the processing efficiency theory. *Cognition and Emotion, 6*(6), 409–434. https://doi.org/10.1080/02699939208409696.

Festinger, L. (1957). *A Theory of Cognitive Dissonance*. Palo Alto: Stanford University Press.

Finucane, M. L., Alhakami, A., Slovic, P., & Johnson, S. M. (2000). The affect heuristic in judgments of risks and benefits. *Journal of Behavioral Decision Making, 13,* 1–17.

Frey, D. (1986). Recent research on selective exposure to information. In L. Berkowitz (Ed.), *Advances in Experimental Social Psychology* (Vol. 19, pp. 41–80). San Diego: Academic Press.

Gigerenzer, G., Todd, P. M., & The ABC Research Group. (1999). *Simple heuristics that make us smart.* New York, NY: Oxford University Press.

Gross, J. J. (1998). The emerging field of emotion regulation: an integrative review. *Review of General Psychology, 2*(3), 271–299. https://doi.org/10.1037/1089-2680.2.3.271.

Johnson, E. J., & Tversky, A. (1983). Affect, generalization, and the perception of risk. *Journal of Personality and Social Psychology, 45*(1), 20–31. https://doi.org/10.1037/0022-3514.45.1.20.

Kahneman, D. (2011). *Thinking, fast and slow*. New York: Farrar, Straus and Giroux.

Kahneman, D., & Frederick, S. (2002). Representativeness revisited: attribute substitution in intuitive judgment. In T. Gilovich, D. W. Griffin, & D. Kahneman (Eds.), *Heuristics and biases: the psychology of intuitive judgment* (pp. 49–81). New York: Cambridge University Press.

Klein, G. A. (1993). A recognition-primed decision (RPD) model of rapid decision making. In *Decision making in action: models and methods* (pp. 138–147). Ablex Pub.

Klein, G. A., Calderwood, R., & Clinton-Cirocco, A. (1986). Rapid decision making on the fire ground. *Proceedings of the Human Factors and Ergonomics Society Annual Meeting, 30*(6), 576–580. https://doi.org/10.1177/154193128603000616.

Lambourne, K., & Tomporowski, P. (2010). The effect of exercise-induced arousal on cognitive task performance: a meta-regression analysis. *Brain Research, 1341,* 12–24. https://doi.org/10.1016/j.brainres.2010.03.091.

Loewenstein, G. F., Weber, E. U., Hsee, C. K., & Welch, N. (2001). Risk as feelings. *Psychological Bulletin, 127*(2), 267–286. https://doi.org/10.1037/0033-2909.127.2.267.

Mano, H. (1992). Judgments under distress: assessing the role of unpleasantness and arousal in judgment formation. *Organizational Behavior and Human Decision Processes, 52*(2), 216–245. https://doi.org/10.1016/0749-5978(92)90036-7.

McCammon, I. (2000). The role of training in recreational avalanche accidents in the United States. In *Proceedings of the international snow science workshop* (pp. 37–45).

McCammon, I. (2004). Heuristic traps in recreational avalanche accidents: evidence and implications. *Avalanche News, 68,* 1–10.

Peters, E., Västfjäll, D., Gärling, T., & Slovic, P. (2006). Affect and decision making: a "hot" topic. *Journal of Behavioral Decision Making, 19,* 79–85.

Raue, M., Kolodziej, R., Lermer, E., & Streicher, B. (2018). Risks seem low while climbing high: shift in risk perception and error rates in the course of indoor climbing activities. *Frontiers in Psychology, 9,* 2383. https://doi.org/10.3389/fpsyg.2018.02383.

Raue, M., & Scholl, S. G. (2018). *The use of heuristics in decision-making under risk and uncertainty.* In M. Raue, E. Lermer, & B. Streicher (Eds.), Psychological perspectives on risk and risk analysis - theory, models and applications. New York, NY: Springer.

Raue, M., Streicher, B., Lermer, E., & Frey, D. (2017). Being active when judging risks: bodily states interfere with accurate risk analysis. *Journal of Risk Research, 20*(4), 445–462. https://doi.org/10.1080/13669877.2015.1057206.

Russell, J. A. (1980). A circumplex model of affect. *Journal of Personality and Social Psychology, 39*(6), 1161–1178. https://doi.org/10.1037/h0077714.

Rydell, R. J., Mackie, D. M., Maitner, A. T., Claypool, H. M., Ryan, M. J., & Smith, E. R. (2008). Arousal, processing, and risk taking: consequences of intergroup anger. *Personality and Social Psychology Bulletin, 34*(8), 1141–1152.

Shiv, B., Loewenstein, G., Bechara, A., Damasio, H., & Damasio, A. R. (2005). Investment behavior and the negative side of emotion. *Psychological Science, 16*(6), 435–439. https://doi.org/10.1111/j.0956-7976.2005.01553.x.

Simon, H. A. (1955). A behavioral model of rational choice. *The Quarterly Journal of Economics, 69*(1), 99–118. https://doi.org/10.2307/1884852.

Slovic, P., & Weber, E. (2002). Perception of risk posed by extreme events. In *Risk management strategies in an uncertain world* (pp. 1–21).

Slovic, P., Finucane, M. L., Peters, E., & MacGregor, D. G. (2004). Risk as analysis and risk as feelings: some thoughts about affect, reason, risk, and rationality. *Risk Analysis, 24*(2), 311–322.

Strack, F., Martin, L. L., & Stepper, S. (1988). Inhibiting and facilitating conditions of the human smile: a nonobtrusive test of the facial feedback hypothesis. *Journal of Personality and Social Psychology, 54*(5), 768–777.

Tversky, A., & Kahneman, D. (1974). Judgment under uncertainty: heuristics and biases. *Science, 185,* 1124–1131.

Williams, L. E., & Bargh, J. A. (2008). Experiencing physical warmth promotes interpersonal warmth. *Science, 322*(5901), 606–607. https://doi.org/10.1126/science.1162548.

Wilson, A. D., & Golonka, S. (2013). Embodied cognition is not what you think it is. *Frontiers in Psychology, 4,* 1–13. https://doi.org/10.3389/fpsyg.2013.00058.

Wilson, M. (2002). Six views of embodied cognition. *Psychonomic Bulletin & Review, 9*(4), 625–636.

Xu, A. J., Zwick, R., & Schwarz, N. (2012). Washing away your (good or bad) luck: physical cleansing affects risk-taking behavior. *Journal of Experimental Psychology: General, 141,* 26–44.

Yuen, K. S. L., & Lee, T. M. C. (2003). Could mood state affect risk-taking decisions? *Journal of Affective Disorders, 75*(1), 11–18. https://doi.org/10.1016/S0165-0327(02)00022-8.

Zhong, C.-B., & Leonardelli, G. J. (2008). Cold and lonely: does social exclusion literally feel cold? *Psychological Science, 19*(9), 838–842. https://doi.org/10.1111/j.1467-9280.2008.02165.x.

Zuckerman, M. (1979). *Sensation seeking: beyond the optimal level of arousal.* Hillsbaum: Lawrence Erlbaum Associates Inc.

Zuckerman, M. (2007). *Sensation seeking and risky behavior.* Washington, D.C.: American Psychological Association.

MIX
Papier aus verantwortungsvollen Quellen
Paper from responsible sources
FSC® C105338
FSC
www.fsc.org

If you have any concerns about our products,
you can contact us on
**ProductSafety@springernature.com**

In case Publisher is established outside the EU,
the EU authorized representative is:
**Springer Nature Customer Service Center GmbH**
**Europaplatz 3, 69115 Heidelberg, Germany**

Printed by Libri Plureos GmbH
in Hamburg, Germany